Isabelle M. Menne

Facing Social Robots

Isabelle M. Menne

Facing Social Robots

Emotional Reactions towards Social Robots

Würzburg
University Press

Dissertation, Julius-Maximilians-Universität Würzburg
Fakultät für Humanwissenschaften, 2019
Gutachter: Prof. Dr. Frank Schwab, Prof. Dr. Gerhild Nieding

Impressum

Julius-Maximilians-Universität Würzburg
Würzburg University Press
Universitätsbibliothek Würzburg
Am Hubland
D-97074 Würzburg
www.wup.uni-wuerzburg.de

©2020 Würzburg University Press
Print on Demand

Coverdesign: Michael Buchta

ISBN 978-3-95826-120-4 (print)
ISBN 978-3-95826-121-1 (online)
DOI 10.25972/WUP-978-3-95826-121-1
URN urn:nbn:de:bvb:20-opus-187131

Preface

No matter how cute today's robots look, no matter how much their virtual faces smile, they will never return your love. The stories of people mourning robots like Jibo, an intelligent domestic worker who announced his own "death" when his servers were to be shut down, are heartwarming. But they also show a way, according to the Associated Press (2019), for marketers to take advantage of people's emotions by programming robots to look more emotionally savvy than they really are. That's the public discussion, but how does the research related to this look like?

Emotional reactions to robots do not only move the surface of the human affective system. They show themselves in the subjective, reportable experience as well as on the motor expressive level. Humans react emotionally to and interact emotionally with robots. This is especially the case when the machines show expressive behavior and somewhat less so when they actually look more like a machine. Even as film recipients, people feel empathy and negative feelings when a social robot is tortured, this is also evident in facial human behaviour.

In this book three elaborated studies are described, which were clearly, stringently and precisely planned. Above all, the robotics aspects were extensively worked on, but immense effort was also put into the human behavioural and experiential side. All studies are comprehensively derived from theory and empiricism and go deeper into the details than many previous human-robot interaction studies (HRI). This work convinces by the conclusiveness of its arguments. At the same time, it remains critical of its own approach. One reads a work on an outstanding level. Successful conclusions and explanations as well as inspiring considerations can be found above all in the summaries and discussions. First and foremost, the work advocates a multi-level or multi-method approach to HRI research – especially when it comes to socio-emotional aspects. The reader can expect a solidly researched, elaborately planned and competently executed work offering new scientific insights into HRI.

Prof. Dr. Frank Schwab

Danksagung

Eine Dissertation wird zwar letztlich von einer Person geschrieben, aber begleitet wird sie von vielen Menschen ohne deren Unterstützung dieses Projekt nicht denkbar gewesen wäre. An dieser Stelle möchte ich all jenen danken, die mich in dieser Zeit gefördert haben, für meine Sorgen und Anliegen immer ein offenes Ohr hatten und mir meine akademische Reise so angenehm wie möglich gestaltet haben.

Zunächst möchte ich mich bei meinen Mentoren bedanken, die diese Dissertation maßgeblich geprägt haben. Mein tiefster Dank gilt meinem Doktorvater, Prof. Dr. Frank Schwab, für seine uneingeschränkte Unterstützung und sein Vertrauen in mich. Seine empathische Art, Kreativität und Offenheit war mir in jeder Hinsicht eine unentbehrliche Hilfe auf dem Weg zur Promotion. Ein besonderer Dank gilt Frau Prof. Dr. Gerhild Nieding, für die Betreuung dieser Arbeit als Zweitgutachterin und Ihre wertvollen Kommentare. Frau Prof. Dr. Ilona Nord bin ich ebenfalls sehr dankbar für Ihre Bereitschaft, drittes Mitglied meines Promotionsgremiums zu sein.

Frau Prof. Dr. Birgit Lugrin und Frau Prof. Dr. Dagmar Unz danke ich dafür, dass sie mir ihre Roboter für meine Experimente zur Verfügung gestellt haben. Den Robotern geht es jetzt wieder gut. Meinen Kolleginnen und Kollegen an der Universität Würzburg danke ich für die Aufmunterungen, unterstützenden Worte und die wunderbare Teamatmosphäre. Besonders danke ich Dorothea Adler für ihre großartige emotionale Unterstützung, Benjamin Lange für seine wertvollen Kommentare zu meinen Studien und Michael Brill für die technische Hilfe. Meinen Bachelorstudentinnen und -studenten danke ich für ihre Hilfe bei der Datenerhebung.

Meinen Eltern gilt tiefster Dank für ihre bedingungslose Unterstützung und ihr Vertrauen in mich. Durch sie wurde diese Arbeit erst möglich. Meiner Schwester danke ich ganz besonders dafür, dass sie immer für mich da ist. Dieser großartige Rückhalt gibt mir sehr viel Kraft und Zuversicht.

Außerdem danke ich meinem Partner Max für seine liebevolle Unterstützung und seine ermutigenden Worte. Ich bin sehr froh ihn an meiner Seite zu haben.

Contents

List of Figures

List of Tables

Abbreviations

FACS   Facial Action Coding System

HRI   Human-Robot Interaction

AU   Action Unit

AUs   Action Units

Abstract

An Army Colonel feels sorry for a robot that defuses landmines on a trial basis and declares the test inhumane (Garreau, 2007). Robots receive military promotions, funerals and medals of honor (Garreau, 2007; Carpenter, 2013). A turtle robot is being developed to teach children to treat robots well (Ackermann, 2018). The humanoid robot Sophia recently became a Saudi Arabian citizen and there are now debates whether robots should have rights (Delcker, 2018). These and similar developments already show the importance of robots and the emotional impact they have. Nevertheless, these emotional reactions seem to take place on a different level, judging by comments in internet forums alone: Most often, emotional reactions towards robots are questioned if not denied at all. In fact, from a purely rational point of view, it is difficult to explain why people should empathize with a mindless machine. However, not only the reports mentioned above but also first scientific studies (e.g. Rosenthal- von der Pütten et al., 2013) bear witness to the emotional influence of robots on humans. Despite the importance of researching emotional reactions towards robots, there are few scientific studies on this subject. In fact, Kappas, Krumhuber and Küster (2013) identified effective testing and evaluation of social reactions towards robots as one of the major challenges of affective Human-Robot Interaction (HRI). According to Scherer (2001; 2005), emotions consist of the coordination and synchronization of different components that are linked to each other. These include motor expression (facial expressions), subjective experience, action tendencies, physiological and cognitive components. To fully capture an emotion, all these components would have to be measured, but such a comprehensive analysis has never been performed (Scherer, 2005). Primarily, questionnaires are used (cf. Bethel & Murphy, 2010) but most of them only capture subjective experiences. Bakeman and Gottman (1997) even state that only about 8% of psychological research is based on behavioral data, although psychology is traditionally defined as the 'study of the mind and behavior' (American Psychological Association, 2018). The measurement of other emotional components is rare. In addition, questionnaires have a number of disadvantages (Austin, Deary, Gibson, Mc-Gregor, Dent, 1998; Fan et al., 2006; Wilcox, 2011). Bethel and Murphy (2010) as well as Arkin and Moshkina (2015) argue for a multi-method approach to achieve a more comprehensive understanding of affective processes in HRI. The main goal of this dissertation is therefore to use a multi-method approach to capture different components of emotions (motor expression, subjective feeling component, action tendencies) and thus contribute to a more complete and pro-found picture of emotional processes towards robots.

To achieve this goal, three experimental studies were conducted with a total of 491 participants. With different levels of 'apparent reality' (Frijda, 2007) and power/control over the situation (cf. Scherer & Ellgring, 2007), the extent to which the intensity and quality of emotional responses to robots change were investigated as well as the influence of other factors (appearance of the robot, emotional expressivity of the robot, treatment of the robot, authority status of the robot).

Experiment 1 was based on videos showing different types of robots (animal-like, anthropomorphic, machine-like) in different situations (friendly treatment of the robot vs. torture treatment) while being either emotionally expressive or not. Self-reports of feelings as well as the motoric-expressive component of emo-tion: facial expressions (cf. Scherer, 2005) were analyzed. The Facial Action Cod-ing System (Ekman, Friesen, & Hager, 2002), the most comprehensive and most widely used method for objectively assessing facial expressions, was utilized for this purpose. Results showed that participants displayed facial expressions (Ac-tion Unit [AU] 12 and AUs associated with positive emotions as well as AU 4 and AUs associated with negative emotions) as well as self-reported feelings in line with the valence of the treatment shown in the videos. Stronger emotional reactions could be observed for emotionally expressive robots than non-expres-sive robots. Most pity, empathy, negative feelings and sadness were reported for the animal-like robot Pleo while watching it being tortured, followed by the an-thropomorphic robot Reeti and least for the machine-like robot Roomba. Most antipathy was attributed to Roomba. The findings are in line with previous re-search (e.g., Krach et al., 2008; Menne & Schwab, 2018; Riek et al., 2009; Rosen-thal-von der Pütten et al., 2013) and show facial expressions' potential for a nat-ural HRI.

Experiment 2 and Experiment 3 transferred Milgram's classic experiments (1963; 1974) on obedience into the context of HRI. Milgram's obedience studies were deemed highly suitable to study the extent of empathy towards a robot in relation to obedience to a robot. Experiment 2 differed from Experiment 3 in the level of 'apparent reality' (Frijda, 2007): based on Milgram (1963), a purely text-based study (Experiment 2) was compared with a live HRI (Experiment 3). While the dependent variables of Experiment 2 consisted of self-reports of emotional feelings and assessments of hypothetical behavior, Experiment 3 measured sub-jective feelings and real behavior (reaction time: duration of hesitation; obedi-ence rate; number of protests; facial expressions) of the participants. Both exper-iments examined the influence of the factors "authority status" (high / low) of the robot giving the orders (Nao) and the emotional expressivity (on / off) of the

robot receiving the punishments (Pleo). The subjective feelings of the participants from Experiment 2 did not differ between the groups. In addition, only few
participants (20.2%) stated that they would definitely punish the "victim" robot.
Milgram (1963) found a similar result. However, the real behavior of participants
in Milgram's laboratory experiment differed from the estimates of hypothetical
behavior of participants to whom Milgram had only described the experiment.
Similarly, comments from participants in Experiment 2 suggest that the scenario
described may have been considered fictitious and that assessments of hypothetical behavior may not provide a realistic picture of real behavior towards robots
in a live interaction. Therefore, another experiment (Experiment 3) was performed with a live interaction with a robot as authority figure (high authority
status vs. low) and another robot as "victim" (emotional expressive vs. non-expressive). Group differences were found in questionnaires on emotional responses. More empathy was shown for the emotionally expressive robot and
more joy and less antipathy was reported than for a non-expressive robot. In addition, facial expressions associated with negative emotions could be observed
while subjects executed Nao's command and punished Pleo. Although subjects
tended to hesitate longer when punishing an emotionally expressive robot and
the order came from a robot with low authority status, this difference did not
reach significance. Furthermore, all but one subject were obedient and punished
Pleo as commanded by the Nao robot. This result stands in stark contrast to the
self-reported hypothetical behavior of the participants from Experiment 2 and
supports the assumption that the assessments of hypothetical behavior in a Human-Robot obedience scenario are not reliable for real behavior in a live HRI.
Situational variables, such as obedience to authorities, even to a robot, seem to
be stronger than empathy for a robot. This finding is in line with previous studies
(e.g. Bartneck & Hu, 2008; Geiskkovitch et al., 2016; Menne, 2017; Slater et al.,
2006), opens up new insights into the influence of robots, but also shows that the
choice of a method to evoke empathy for a robot is not a trivial matter (cf.
Geiskkovitch et al., 2016; cf. Milgram, 1965). Overall, the results support the assumption that emotional reactions to robots are profound and manifest both at
the subjective level and in the motor component. Humans react emotionally to a
robot that is emotionally expressive and looks less like a machine. They feel empathy and negative feelings when a robot is abused and these emotional reactions
are reflected in facial expressions. In addition, people's assessments of their own
hypothetical behavior differ from their actual behavior, which is why video-based
or live interactions are recommended for analyzing real behavioral responses.
The arrival of social robots in society leads to unprecedented questions and this
dissertation provides a first step towards understanding these new challenges.

Zusammenfassung

Ein Army Colonel empfindet Mitleid mit einem Roboter, der versuchsweise Landminen entschärft und deklariert den Test als inhuman (Garreau, 2007). Roboter bekommen militärische Beförderungen, Beerdigungen und Ehrenmedaillen (Garreau, 2007; Carpenter, 2013). Ein Schildkrötenroboter wird entwickelt, um Kindern beizubringen, Roboter gut zu behandeln (Ackermann, 2018). Der humanoide Roboter Sophia wurde erst kürzlich Saudi-Arabischer Staatsbürger und es gibt bereits Debatten, ob Roboter Rechte bekommen sollen (Delcker, 2018). Diese und ähnliche Entwicklungen zeigen schon jetzt die Bedeutsamkeit von Robotern und die emotionale Wirkung die diese auslösen. Dennoch scheinen sich diese emotionalen Reaktionen auf einer anderen Ebene abzuspielen, gemessen an Kommentaren in Internetforen. Dort ist oftmals die Rede davon, wieso jemand überhaupt emotional auf einen Roboter reagieren kann. Tatsächlich ist es, rein rational gesehen, schwierig zu erklären, warum Menschen mit einer leblosen (‚mindless') Maschine mitfühlen sollten. Und dennoch zeugen nicht nur oben genannte Berichte, sondern auch erste wissenschaftliche Studien (z.B. Rosenthal- von der Pütten et al., 2013) von dem emotionalen Einfluss den Roboter auf Menschen haben können. Trotz der Bedeutsamkeit der Erforschung emotionaler Reaktionen auf Roboter existieren bislang wenige wissenschaftliche Studien hierzu. Tatsächlich identifizierten Kappas, Krumhuber und Küster (2013) die systematische Analyse und Evaluation sozialer Reaktionen auf Roboter als eine der größten Herausforderungen der affektiven Mensch-Roboter Interaktion. Nach Scherer (2001; 2005) bestehen Emotionen aus der Koordination und Synchronisation verschiedener Komponenten, die miteinander verknüpft sind. Motorischer Ausdruck (Mimik), subjektives Erleben, Handlungstendenzen, physiologische und kognitive Komponenten gehören hierzu. Um eine Emotion vollständig zu erfassen, müssten all diese Komponenten gemessen werden, jedoch wurde eine solch umfassende Analyse bisher noch nie durchgeführt (Scherer, 2005). Hauptsächlich werden Fragebögen eingesetzt (vgl. Bethel & Murphy, 2010), die allerdings meist nur das subjektive Erleben abfragen. Bakeman und Gottman (1997) geben sogar an, dass nur etwa 8% der psychologischen Forschung auf Verhaltensdaten basiert, obwohl die Psychologie traditionell als das ‚Studium von Psyche und Verhalten' (American Psychological Association, 2018) definiert wird. Die Messung anderer Emotionskomponenten ist selten. Zudem sind Fragebögen mit einer Reihe von Nachteilen behaftet (Austin, Deary, Gibson, McGregor, Dent, 1998; Fan et al., 2006; Wilcox, 2011). Bethel und Murphy (2010) als auch Arkin und Moshkina (2015) plädieren für einen Multi-Methodenansatz um ein umfassenderes Verständnis von affektiven Prozessen in der Mensch-Roboter Interaktion zu erlangen. Das Hauptziel der vorliegenden Dissertation ist es daher, mithilfe eines Multi-Methodenansatzes verschiedene Komponenten von Emotionen (motorischer Ausdruck, subjektive Gefühlskomponente, Handlungstendenzen) zu erfassen und so zu einem vollständigeren und tiefgreifenderem Bild emotionaler Prozesse auf Roboter beizutragen.

Um dieses Ziel zu erreichen, wurden drei experimentelle Studien mit insgesamt 491 Teilnehmern durchgeführt. Mit unterschiedlichen Ebenen der „apparent reality" (Frijda, 2007) sowie Macht / Kontrolle über die Situation (vgl. Scherer & Ellgring, 2007) wurde untersucht, inwiefern sich Intensität und Qualität emotionaler Reaktionen auf Roboter ändern und welche weiteren Faktoren (Aussehen des Roboters, emotionale Expressivität des Roboters, Behandlung des Roboters, Autoritätsstatus des Roboters) Einfluss ausüben.

Experiment 1 basierte auf Videos, die verschiedene Arten von Robotern (tierähnlich, anthropomorph, maschinenartig), die entweder emotional expressiv waren oder nicht (an / aus) in verschiedenen Situationen (freundliche Behandlung des Roboters vs. Misshandlung) zeigten. Fragebögen über selbstberichtete Gefühle und die motorisch-expressive Komponente von Emotionen: Mimik (vgl. Scherer, 2005) wurden analysiert. Das Facial Action Coding System (Ekman, Friesen, & Hager, 2002), die umfassendste und am weitesten verbreitete Methode zur objektiven Untersuchung von Mimik, wurde hierfür verwendet. Die Ergebnisse zeigten, dass die Probanden Gesichtsausdrücke (Action Unit [AU] 12 und AUs, die mit positiven Emotionen assoziiert sind, sowie AU 4 und AUs, die mit negativen Emotionen assoziiert sind) sowie selbstberichtete Gefühle in Übereinstimmung mit der Valenz der in den Videos gezeigten Behandlung zeigten. Bei emotional expressiven Robotern konnten stärkere emotionale Reaktionen beobachtet werden als bei nicht-expressiven Robotern. Der tierähnliche Roboter Pleo erfuhr in der Misshandlungs-Bedingung am meisten Mitleid, Empathie, negative Gefühle und Traurigkeit, gefolgt vom anthropomorphen Roboter Reeti und am wenigsten für den maschinenartigen Roboter Roomba. Roomba wurde am meisten Antipathie zugeschrieben. Die Ergebnisse knüpfen an frühere Forschungen an (z.B. Krach et al., 2008; Menne & Schwab, 2018; Riek et al., 2009; Rosenthalvon der Pütten et al., 2013) und zeigen das Potenzial der Mimik für eine natürliche Mensch-Roboter Interaktion.

Experiment 2 und Experiment 3 übertrugen die klassischen Experimente von Milgram (1963; 1974) zum Thema Gehorsam in den Kontext der Mensch-Roboter Interaktion. Die Gehorsamkeitsstudien von Milgram wurden als sehr geeignet erachtet, um das Ausmaß der Empathie gegenüber einem Roboter im Verhältnis zum Gehorsam gegenüber einem Roboter zu untersuchen. Experiment 2 unterschied sich von Experiment 3 in der Ebene der „apparent reality" (Frijda,

2007): in Anlehnung an Milgram (1963) wurde eine rein text-basierte Studie (Experiment 2) einer live Mensch-Roboter Interaktion (Experiment 3) gegenübergestellt. Während die abhängigen Variablen von Experiment 2 aus den Selbstberichten emotionaler Gefühle sowie Einschätzungen des hypothetischen Verhaltens bestand, erfasste Experiment 3 subjektive Gefühle sowie reales Verhalten (Reaktionszeit: Dauer des Zögerns; Gehorsamkeitsrate; Anzahl der Proteste; Mimik) der Teilnehmer. Beide Experimente untersuchten den Einfluss der Faktoren „Autoritätsstatus" (hoch / niedrig) des Roboters, der die Befehle erteilt (Nao) und die emotionale Expressivität (an / aus) des Roboters, der die Strafen erhält (Pleo). Die subjektiven Gefühle der Teilnehmer aus Experiment 2 unterschieden sich zwischen den Gruppen nicht. Darüber hinaus gaben nur wenige Teilnehmer (20.2%) an, dass sie den „Opfer"-Roboter definitiv bestrafen würden. Ein ähnliches Ergebnis fand auch Milgram (1963). Das reale Verhalten von Versuchsteilnehmern in Milgrams' Labor-Experiment unterschied sich jedoch von Einschätzungen hypothetischen Verhaltens von Teilnehmern, denen Milgram das Experiment nur beschrieben hatte. Ebenso lassen Kommentare von Teilnehmern aus Experiment 2 darauf schließen, dass das beschriebene Szenario möglicherweise als fiktiv eingestuft wurde und Einschätzungen von hypothetischem Verhalten daher kein realistisches Bild realen Verhaltens gegenüber Roboter in einer live Interaktion zeichnen können. Daher wurde ein weiteres Experiment (Experiment 3) mit einer Live Interaktion mit einem Roboter als Autoritätsfigur (hoher Autoritätsstatus vs. niedriger) und einem weiteren Roboter als „Opfer" (emotional expressiv vs. nicht expressiv) durchgeführt. Es wurden Gruppenunterschiede in Fragebögen über emotionale Reaktionen gefunden. Dem emotional expressiven Roboter wurde mehr Empathie entgegengebracht und es wurde mehr Freude und weniger Antipathie berichtet als gegenüber einem nicht-expressiven Roboter. Außerdem konnten Gesichtsausdrücke beobachtet werden, die mit negativen Emotionen assoziiert sind während Probanden Nao's Befehl ausführten und Pleo bestraften. Obwohl Probanden tendenziell länger zögerten, wenn sie einen emotional expressiven Roboter bestrafen sollten und der Befehl von einem Roboter mit niedrigem Autoritätsstatus kam, wurde dieser Unterschied nicht signifikant. Zudem waren alle bis auf einen Probanden gehorsam und bestraften Pleo, wie vom Nao Roboter befohlen. Dieses Ergebnis steht in starkem Gegensatz zu dem selbstberichteten hypothetischen Verhalten der Teilnehmer aus Experiment 2 und unterstützt die Annahme, dass die Einschätzungen von hypothetischem Verhalten in einem Mensch-Roboter-Gehorsamkeitsszenario nicht zuverlässig sind für echtes Verhalten in einer live Mensch-Roboter Interaktion. Situative Variablen, wie z.B. der Gehorsam gegenüber Autoritäten, sogar gegenüber einem Roboter, scheinen stärker zu sein als Empathie für einen Roboter. Dieser Befund knüpft an andere Studien an (z.B. Bartneck & Hu, 2008; Geiskkovitch et al., 2016;

Menne, 2017; Slater et al., 2006), eröffnet neue Erkenntnisse zum Einfluss von Robotern, zeigt aber auch auf, dass die Wahl einer Methode um Empathie für einen Roboter zu evozieren eine nicht triviale Angelegenheit ist (vgl. Geiskko-vitch et al., 2016; vgl. Milgram, 1965). Insgesamt stützen die Ergebnisse die An-nahme, dass die emotionalen Reaktionen auf Roboter tiefgreifend sind und sich sowohl auf der subjektiven Ebene als auch in der motorischen Komponente zei-gen. Menschen reagieren emotional auf einen Roboter, der emotional expressiv ist und eher weniger wie eine Maschine aussieht. Sie empfinden Empathie und negative Gefühle, wenn ein Roboter misshandelt wird und diese emotionalen Re-aktionen spiegeln sich in der Mimik. Darüber hinaus unterscheiden sich die Ein-schätzungen von Menschen über ihr eigenes hypothetisches Verhalten von ih-rem tatsächlichen Verhalten, weshalb videobasierte oder live Interaktionen zur Analyse realer Verhaltensreaktionen empfohlen wird. Die Ankunft sozialer Ro-boter in der Gesellschaft führt zu nie dagewesenen Fragen und diese Dissertation liefert einen ersten Schritt zum Verständnis dieser neuen Herausforderungen.

1 Introduction

> At the Yuma Test Grounds in Arizona, the autonomous robot, 5 feet long
> and modeled on a stick-insect, strutted out for a livefire test and worked
> beautifully, he [Mark Tilden, the creator of the robot] says. Every time it
> found a mine, blew it up and lost a limb, it picked itself up and readjusted
> to move forward on its remaining legs, continuing to clear a path through
> the minefield. Finally it was down to one leg. Still, it pulled itself forward.
> Tilden was ecstatic. The machine was working splendidly.
> The human in command of the exercise, however – an Army colonel –
> blew a fuse.
> The colonel ordered the test stopped. Why? asked Tilden. What's wrong?
> The colonel just could not stand the pathos of watching the burned,
> scarred and crippled machine drag itself forward on its last leg.
> This test, he charged, was inhumane. (Garreau, 2007, para. 2-7)

The robot was equipped with several legs to detonate land mines and in the process, each time one of the robot's legs was destroyed. Even though the robot was
specifically built for this purpose, watching a mindless, lifeless machine getting
ripped to shreds evoked feelings of empathy in the human. This example is by far
not the only (anecdotal) evidence of emotional reactions towards robots. Robots
have been awarded "battlefield promotions", "purple hearts" (Garreau, 2007,
para. 10), funerals and medals of honor (Carpenter, 2013). A recent press release
stated that "participants rather save a robot than a human" (Bruns, 2019). Even
though the corresponding journal article that the online press release refers to
phrases it more carefully: "when people attribute affective states to robots, they
are less likely to sacrifice them in order to save humans" (Nijssen, Müller, van
Baaren, & Paulus, 2019), the message is clear. And much controversy surrounds
it judging from user comments alone. One user for example wrote[1] "Why should
you want to save a robot? It can easily be repaired and its database probably exists
on the net anyway. Downloaded and copied into the next sheet metal body it is
again ready for action" (Wraithsong, 2019). That this might not be so easily done
is also shown by anecdotal evidence of a veteran explosives technician whose military robot was torn to pieces. The repairer advised him to get a new robot, but
he refused, instead wanting his "Scooby-Doo" back (Garreau, 2007). Similar reactions have been reported for the vacuum cleaning robot Roomba: It is called
"Roomba baby", gets introduced to the owner's parents and its human owners

---

[1] Translated by author. Original comment in german.

actually clean for the robot so that it can rest (Sung, Guo, Grinter, & Christensen, 2007). Stieler (2019) even advocates to "love your machines!²" and to "stop the violence against robots!³" (para. 6). Indeed, websites have been installed to draw attention to the abuse of robots and prevent it (e.g., stoprobotabuse.com). A turtle robot has been designed to teach children to treat robots well (Ackermann, 2018). Robots have even entered the political level: the humanoid robot Sophia has "become a Saudi Arabian citizen, was given a title from the United Nations and opened the Munich Security Conference" (Delcker, 2018, "Beyond Sophia", para. 5). Furthermore, there are debates on whether robots should have rights (Delcker, 2018, para. 3).

Robots are starting to enter people's homes and become part of their everyday life. They come in the shape of lawn mowers, vacuum cleaner robots or entertainment robots. The arrival of robots in people's private lives leads to new challenges and questions, such as: how close should social robots get? Is it morally acceptable to mistreat them, tear them apart and sell them without feeling bad? Even if it looks like a human, acts as though it's alive and expresses realistic emotions? In other words, how deeply ingrained is people's propensity disposition to treat artificial social entities as humans? The answer to these questions relies on understanding the determinants and mechanisms of emotional reactions towards robots.

Despite its huge impact on society, systematic research on emotional reactions towards robots remains scarce: effective testing and evaluation of social responses robots evoke is a major challenge in HRI (Kappas, Krumhuber, & Küster, 2013; Eyssel, 2017).

A recent press release of the german scientific board (2018) called upon psychology to be more open to address relevant issues in society as well as to contribute to dealing with key societal challenges: "The need for insights into the phenomena of human experience and behavior is greater than ever and is also growing in new fields, such as those associated with the buzz words 'user experience' or 'industry 4.0' "⁴ (German Scientific Board, 2018).

This doctoral dissertation aims to contribute to a current and future societal challenge: the emotional impact of the arrival of a "new species": social robots. Several aspects are considered important to reach this goal: first, a multi-method approach is chosen by analyzing a) spontaneous facial expressions as indicators of emotional processes, b) self-reports of individual's subjective emotional experiences and c) further behavioral measures (obedience rate, hesitation time). This

---

² Translated by author. Original comment in german.
³ Translated by author. Original comment in german.
⁴ Translated by author. Original statement in german.

does not only provide a more complete picture of emotions as a multi-level phenomenon. It also meets the needs of effective testing and evaluation, which is identified as one of the major challenges in affective human-robot interaction (HRI) (Kappas et al., 2013; see also Eyssel, 2017). Second, it is explored how people would respond to the intense dilemma of empathy versus obedience in the context of HRI by using a variation of the obedience scenario by Milgram (1963) in vivo (laboratory experiment) compared to in sensu (web-based questionnaire). By using these factors, a more complete and profound understanding of the emotional impact of social robots can be gained.

2 Scope

The aim of this doctoral dissertation is to investigate the profoundness of emotional reactions towards social robots. Three studies have been conducted to explore the influence of different levels of 'apparent reality' (Frijda, 2007) on users' affective responses in situations rising in emotional intensity. According to Frijda (2007), pictures or events actually seen generally have a greater impact than symbolic information. This phenomenon is also discussed as the 'vividness effect' (Fiske & Taylor, 1984). Hence, the key challenge was to discover whether there were differences in emotional reactions between a) simply watching a robot being mistreated (Experiment 1), b) to imagine mistreating a robot (Experiment 2) and c) actually having to mistreat a robot oneself (Experiment 3). The impact of both, the different levels of apparent reality as well as the change in role for the participant is investigated. Whereas participants simply react towards the display of violent behavior in Experiment 1, the costs of allowing oneself emotional feelings towards a robot are considerably higher: in Experiment 2 and 3, users are put in an agentic role of being able to decide whether to obey an authority or to follow their emotions and not hurt anyone. Hence, by giving users a choice to stop the mistreatment of a robot, the psychological situation becomes more intense. But there might not only be a change in emotion quanitity but also in emotion quality: According to the Component Process Model (Scherer, 1984; 2001), there are four major appraisal checks (relevance, implications, coping potential, and norm compatibility) for an adaptive reaction to an event. Regarding coping potential, if the event is evaluated as low in power and control (low coping potential), other Action Units (AUs) and emotions are triggered than if the event is evaluated as high in power and control (high coping potential). While Experiment 1, where participants are not given a choice to stop the mistreatment of the robot (low power/control), emotions such as fear or sadness as well as different AUs[5] might occur (Scherer & Ellgring, 2007), whereas Experiment 2 and 3 both put the user in an agentic role (high power/control) where emotions like anger, joy, and disgust and different AUs[6] are predicted to appear according to Scherer & Ellgring (2007). Furthermore, the decision becomes harder. Participants are put in a moral dilemma of having to choose between obedience to authority and empathy with the victim (not to hurt the robot). Will feelings of empathy be able to overcome strong social pressure? Does it depend on the robot's authority status? The

---

[5] AUs 20, 26, 27
[6] 23+25; 17+23; 6+17+24

arrival of social robots presents an unprecedentend societal challenge. The purpose of this dissertation is to test the ground for the profoundness of emotional reactions and obedience towards robots.

3 Theoretical Background

This section illustrates the main theoretical conceptions that form the basis for the empirical part of this dissertation. There are six main parts: social robots, emotions, empathy, facial expressions, measurement of emotions, and obedience. To understand the concept of social robots, which is used in the following sections, the first part of the theoretical background defines what is meant by calling a robot social and explains human-robot interaction (HRI) and related terms. Furthermore, research on emotionally expressive robots is illustrated. The next section concentrates on emotions in psychology and HRI. As emotions play a major role for this doctoral dissertation, different conceptualizations of emotions as well as distinctions between different affective concepts are given. Next, emotion theories as well as research on emotions both in psychology as well as in HRI are presented. Empathy and facial expressions, both closely related to emotions are the next two topics. There is also a special focus on these concepts in the light of HRI. Exploring the relation between obedience and empathy towards robots, another topic is thus obedience research in psychology. Milgram's studies on obedience are presented as well as their consequences (e.g. ethical considerations). Furthermore, research on obedience in psychology as well as in HRI is illustrated.

Even though psychology is defined as "the study of the mind and behavior" (American Psychological Association, 2018) only about 8% of psychological research is based on behavioral data (e.g., Bakeman & Gottman, 1997). Hence, a major focus of this doctoral dissertation was to combine observational methods based on behavioral data with self-report measurements. That way, a broader range of different levels of affective phenomena can be captured. Hence, as Experiment 1 and 3 used this approach, different measurement methods of emotions are presented and evaluated in a concluding remark. Additionally, advantages and disadvantages of online research (focusing on self-reports) in contrast to laboratory research (focusing on behavioral data) are introduced.

3.1 Social Robots

In the following, to understand what is meant by a social robot, a definition of this term as well as of HRI will be given. Furthermore, this section illustrates how features of social robots, such as being emotionally expressive, affect humans' perceptions and actions.

3.1.1 Social Robots and HRI

There is no consistent definition of what a social robot actually is, but a few definitions with roughly similar criteria exist. Dautenhahn and Billard (1999), for example, describe the term as follows:

> Social robots are embodied agents that are part of a heterogeneous group: a society of robots or humans. They are able to recognize each other and engage in social interactions, they possess histories (perceive and interpret the world in terms of their own experience), and they explicitly communicate with and learn from each other.

Fong, Nourbakhsh, and Dautenhahn (2003) further use the term "socially interactive robots" (p. 3) to emphasize the robot's ability for social interaction. The characteristics those robots exhibit include, among others, the ability to express and perceive emotions, establish and maintain social relationships and use natural cues (Fong, Nourbakhsh et al., 2003).

Leite, Martinho, and Paiva (2013), in their survey about social robots for long-term interaction, studied social robots as those "designed to socially interact with people or to evoke social responses from them" (p. 291).

According to Feil-Seifer and Matarić (2009), research on social robots is a subcategory of HRI since it focuses on social interaction. HRI itself is an interdisciplinary field concerned with the "analysis, design, modelling, implementation and evaluation of robots for human use" (Fong, Thorpe, & Baur, 2003).

In the following, the term HRI and social robotics will thus be used synonymously.

A socially interactive robot is, among others, required to be able to send readable signals to its human interaction partner and exhibit competent behavior, conveying attention and intentionality (Fong, Nourbakhsh et al., 2003). Heider and Simmel (1944) showed participants animations of differently shaped objects (e.g., shaped as a triangle) and found that they attributed intentions and even personalities to the moving shapes. Indeed, it has been argued that, in order to survive, an organism has to be able to decode information based on another beings' movements (Johansson, von Hofsten, & Jansson, 1980; Pavlova, Krägeloh-Mann, Birbaumer, & Sokolov, 2002; Pollick, Paterson, Bruderlin, & Sanford, 2001; Troje, 2003), gestures and expressions (Blake & Shiffrar, 2007; Troje, 2003). Social cues such as facial expressions, vocal tone or body postures help provide more information about the interaction partner and thus reduce ambiguity which is one of the most important effects of social cues (e.g., Sheth et al., 2011).

According to Dennett (1987), there are three strategies to understand and predict behavior: the physical stance, the design stance and the intentional stance.

For simple systems, the first two stances are sufficient, since predictions can be made based on either physical features or design and functionality features of artificial entities. For more complex systems like humans, an intentional stance has to be adopted since physical characteristics or design features are invalid. Instead, humans tend to rely on beliefs and desires. Fong, Nourbakhsh, et al. (2003) therefore argue that a robot should be able to show intentionality in order to interact socially. This has mostly been realized by using behavioral cues. Kismet, for example, one of the first 'social' robots, conveyed intentionality by using facial expressions (Breazeal & Scassellati, 1999). Vlachos and Schärfe (2015) also state that a robot using facial emotional expressions conveys intentionality by providing "information to its surrounding environment about its most probable following action" (p. 746).

Indeed, it could be shown that robots exhibiting nonverbal behavior have several positive effects, such as they were rated more positively (e.g., Leite, Martinho, Pereira, & Paiva, 2008; Leite et al., 2010; Pereira, Leite, Mascarenhas, Martinho, & Paiva, 2011; Salem, Eyssel, Rohlfing, Kopp, & Joublin, 2011) and even influences on user's emotional states have been suggested (Xu, Broekens, Hindriks, & Neerincx, 2014).

3.1.2 Emotional Expressiveness in Robots

There is a wider variety of research on the effect of a robot's emotional expression on humans than there is on emotional reactions of humans towards a robot. In the following, research on the effect of a robot's verbal and nonverbal expression is highlighted as a major part of this dissertation is concerned with the influence of a robot's emotional expressivity on emotional reactions.

Research not explicitly focusing on a comparison between affective and non-affective robots has shown that people are prone to react emotionally to a robot's emotional behavior (e.g., Menne & Schwab, 2018; Rosenthal-von der Pütten, Krämer, Hoffmann, Sobieraj, & Eimler, 2013; Suzuki et al., 2015).

Since 60-65% of interpersonal communication takes place via nonverbal behavior (Burgoon, Guerrero, & Manusov, 2011), it can be used for conveying intention and proofs to be more effective in collaborative tasks (Breazeal, Kidd, Thomaz, Hoffman, & Berlin, 2005; Mutlu, Yamaoka, Kanda, Ishiguro, & Hagita, 2009). Furthermore, robots using nonverbal behavior are able to convey emotions (Embgen et al., 2012; Häring, Bee, & André, 2011).

As facial expression is the first ability towards social robots (Littlewort, 2004) and thus very important for (humanoid) robots (Park, Lee, & Chung, 2015; Trovato & Takanishi, 2015), they have been frequently used. Most studies in social robotics use facial expressions for conveying emotions (e.g., Breazeal, 2002a; Cañamero & Fredslund, 2000; Fong, Nourbakhsh et al., 2003; Kirby, Forlizzi, & Simmons, 2010; Mirnig et al., 2015; see Calvo, D'Mello, Gratch, & Kappas, 2015, for an overview). For instance, in a study with an emotional expressive character compared to a non-expressive character, users preferred the interaction with the emotional expressive robotic character (Bartneck, 2003).

For robots unable to physically express facial emotions typically use body movements, posture, orientation, color, and sound as a type of nonverbal communication (e.g., Beck, Cañamero, & Bard, 2010; Li & Chignell, 2011; Bethel & Murphy, 2010b). For example, a robot orienting towards a human indicates its perceived attentiveness and affection (e.g., Fong, Nourbakhsh et al., 2003; Bruce, Nourbakhsh, & Simmons, 2002; Dautenhahn et al., 2006). In a study comparing the robot dog AIBO with a stuffed animal, young children were observed to engage in exploratory and apprehensive behavior with the robot dog and mistreated the stuffed dog more often. However, no difference in children's reported evaluations of AIBO and the stuffed dog was found. The authors conclude that social robots challenge the traditional ontological categories like differentiation between animate and inanimate because terms like autonomous, adaptive and embodied are, to some extent, also valid for social robots (Kahn, Friedman, Pérez-Granados, & Freier, 2006).

Examples for affective expressions via voice modulations can be found in Scheutz, Schermerhorn, & Kramer, 2006 who showed that participants worked more efficiently as the robot's anxiety (via speech rate and pitch) increased. In another task performance study, Moshkina (2012) reported that participants responded earlier and moved faster in response to an affective robot than a non-affective robot in a mock up search and rescue setting. The impression of affect was created by using nonverbal (e.g., head lowered) and verbal signals (e.g., higher pitched voice and faster) (Moshkina, 2012).

3.2 Emotions

To get an understanding of what emotions are, literature provides an overwhelming large array of emotion definitions, concepts, models, measurements and emotion terms. There are different approaches to the topic of emotions and the following sections will first illustrate difficulties in defining emotions and present a componential approach. Then, an overview of relevant emotion theories is given. Related emotion research in HRI highlights the challenges of emotion research in the field of social robotics.

3.2.1 Emotions in General

"What is an emotion?" James (1884) was not the first to ask this question but the title of this essay already suggests the question is not easily answered. Indeed, there is a wide variety of different concepts of emotions and Kleinginna and Kleinginna (1981) systematically reviewed existing literature, resulting in a list of 92 definitions and theoretical explanations of emotions. Izard (2010), in an effort to systematically integrate all emotional phenomena into one conceptualization came to the conclusion that researchers should provide operational definitions of their research object. Despite the grand diversity of emotion models, many authors share the view of emotions as a multi-level phenomenon. Usually, five components of emotions are reported (Moors, 2009): a cognitive, neurophysiological, motivational, motoric-expressive and subjective feeling component (see Table 1).

Some authors refuse to adopt a cognitive component to an emotional reaction (Zajonc, 1980). This view is supported by neuroscientific findings: Correlations between emotional responses and subcortical activity suggest emotions can emerge automatically and unconsciously (Cacioppo, Gardner, & Berntson, 1999). However, there are also contradictory findings (e.g., Rolls, Hornak, Wade,

Table 1. Emotions as a multi-level phenomenon (adapted from Scherer, 1984; 2005; see also, e.g., Merten, 2003)

Component	Function	Organismic subsystem
Cognitive (appraisal)	Evaluation of objects and events	Information processing system
Neurophysiological (bodily symptoms)	System regulation	Support system
Motivational (action tendencies)	Action preparation	Executive system
Motoric-expressive (facial and vocal expression)	Communication of emotional reaction	Action system
Subjective feeling (emotional experience)	Reflection, control	Monitor system

& McGrath, 1994). The legitimation of the cognitive component is still treated-controversially (cf. Damasio, 2003). In a compromise, it is supposed that there are conscious and unconscious associations between cognitive and affective parts of emotions. Scherer (1984, 2001) managed to integrate these concerns into his Component Process Model (described in section 3.2.4.3). Bischof (1989) points out that – contrary to a common view ("due to deep-rooted misunderstanding of basic biological principles", p. 204) – emotions are not primarily obstructions to rational thought and but have developed in the course of evolution due to a selective advantage. According to Bischof (1989), emotions require the processing of information, a cognitive process, hence feelings themselves are cognitive. For further details on the scientific debate please refer to Bischof (1989); cf. Dörner (1989), Scherer (1989), Schneider (1989), and Zajonc (1989).

Shiota and Kalat (2012) conclude: "Whether you think appraisal is necessary for emotion depends in part on which definition you are using" (p. 357).

3.2.2 Distinctions Between Different Affective Concepts

Confusion often arises when trying to distinguish emotion from affect and mood. Several authors have proposed that moods are less intense, longer lasting and have a less clear object focus (see Rosenberg, 1998, for a discussion). Moods are thought of as lacking a specific object focus but with a longer duration than emotions (Gross, 2010; see also Davidson, Scherer, & Goldsmith, 2009 for a review). Gross (2010) proposes mood and emotion as subcategories of affect which in turn refers to "valenced (good versus bad) states" (p. 212). Admitting that "the whole lexicon of emotion-related terms is in a bit of a jumble" (Gross, 2010, p. 212), the author speaks of an empirical challenge for understanding and comparing the emotions authors are studying. Due to the conceptual ambiguity, Pekrun (2006) views moods and emotions as "parts of one and the same multi-dimensional space of emotions, rather than distinct categories" (p. 316). Several researchers thus refrain from distinguishing emotions and moods at all and instead use the term "affect" as the main category for all emotional experiences (e.g., Gross, 2010; Walker-Andrews, 2008). Furthermore, "affect" and "emotion" are often used interchangeably in English-speaking regions (e.g., Gross, 1998).

Hence, in this dissertation, the terms "affects" and "emotions" are used synonymously, encompassing all types of affective phenomena.

3.2.3 Induction of Emotions

The three most common methods for inducing emotions in participants in laboratory settings are (1) recalls of emotional incidents in participants' past (e.g. Bless et al., 1996; Ekman, Levenson, & Friesen, 1983) (2) reading about an emotional story (e.g. Griskevicius, Shiota, & Neufeld, 2010; Keltner, Ellsworth, & Edwards, 1993) or (3) viewing short emotional film clips (e.g. Gross & Levenson, 1995; Maner et al., 2005; Papousek, Schulter, & Lang, 2009). Lazarus (1991) argued that emotions induced by films are real emotions. For a discussion on the use of films in HRI see, e.g., Woods, Walters, Koay, and Dautenhahn (2006).

3.2.4 Emotion Theories

Studying emotions is a complex task as there is a vast variety of emotion theories, lacking an integrative, comprehensive emotion theory. However, four most prevalent current theories in emotion research can be identified: basic emotion theory (e.g., viewing emotions in distinct emotion terms, such as sadness, happiness; Ekman, 1992), dimensional emotions approach (e.g., positive and negative affect [valence] or arousal; Watson & Tellegen, 1985), appraisal theories (Scherer, 2001) and constructivist emotions approach (i.e. emotions are cognitively constructed concepts; Barrett, 2006; Lindquist, 2013). The first two approaches to emotion are also widely used in HRI and further technical applications (Calvo & D'Mello, 2010) (see section 3.2.5). Furthermore, since this dissertation focuses on facial emotional expressions including dimensional aspects, the latter approach will not be discussed further (for current views on the constructivist emotions approach see Barrett, 2014; Mesquita & Boiger, 2014).

3.2.4.1 Basic Emotions

The theory of basic emotions is still highly popular among researchers as it "has been and remains the major program for scientific research on emotion" (Russell, Rosenberg, & Lewis, 2011, p. 363). The basic emotion theory belongs to the discrete (categorical) approaches. Some theorists (Ekman, 1992; Izard, 1992; Plutchik, 1980) have postulated a set of basic emotion categories where each emotion (e.g. sadness, happiness) corresponds to a unique pattern in experience, physiology, and behavior (see also Ekman, 2005). Especially facial expressions are linked to each of those emotion categories (Ekman, Friesen, & Ellsworth, 2013). Based on studies in archaic cultures in Papua New Guinea (Ekman & Friesen, 1971), it could be proved that there is a limited set of emotions that are "basic" as in being inherently and culturally invariant: happiness, anger, sadness, disgust, fear, surprise and contempt (Ekman, 2013). Ekman and Cordaro (2011, p. 365) described 13 characteristics of basic emotions that allow differentiation of diverse emotions from one another as well as differentiation from other "affective phenomena, such as moods or emotional traits" (Ekman, 2005, p. 47) (see Table 2).

According to Ekman and Cordaro (2011) the listed criteria can be used as a guide to whether an emotion can be classified as a basic emotion.

The concept of basic emotions is primarily associated with the analysis of facial expressions. Since the analysis of facial expressions is one of the main aspects in this dissertation, evidence indicating facial expressions are associated with emotions is reported in an own section (section 3.4).

Table 2. Characteristics which distinguish basic emotions from each other and other affective phenomena (Ekman & Cordaro, 2011, p. 365)

1	Distinctive universal signals
2	Distinctive physiology
3	Automatic appraisal
4	Distinctive universals in antecedent events
5	Presence in other primates
6	Capable of quick onset
7	Can be of brief duration
8	Unbidden occurrence
9	Distinctive thoughts, memories, and images
10	Distinctive subjective experience
11	Refractory period filters information available to what supports the emotion.
12	Target of emotion unconstrained
13	The emotion can be enacted in either a constructive or destructive fashion

3.2.4.2 Dimensional Theories

Dimensional approaches agree that there are a limited number of dimensions underlying emotion categories. Valence, arousal and approach-avoidance are among the most commonly assumed dimensions (Davidson, 2005; Lang, Bradley, & Cuthbert, 1997; Russell & Barrett, 1999; Schneirla, 1959; Watson, Wiese, Vaidya, & Tellegen, 1999). The valence dimension contains emotional states of pleasure (positive affect) and displeasure (negative affect). Whereas Tellegen, Watson, and Clark (1999) as well as Larsen, McGraw, and Cacioppo (2001) state that positive and negative emotions are relatively independent of each other, others (e.g. Russell, 1980) have taken the view that they are inversely related. Some also argue that approach and avoidance are the same as positive and negative affect, respectively (Watson et al., 1999). Even though more recent research suggests that basic emotional states are reflected in those two dimensions, more dimensions are required for a general approach (Fontaine, Scherer, Roesch, & Ellsworth, 2007). This presents also the main drawback of dimensional models: it is still disputed how severely dimensional models restrict the range of emotional states.

Smith and Ellsworth (1985) as well as Haidt and Keltner (1999) point out that each discrete emotion represents a combination of several dimensions which makes it possible to reconcile dimensional and discrete perspectives to some extent.

3.2.4.3 Appraisal Theories

Contrary to basic emotion theories, appraisal theories of emotion postulate appraisal processes precede emotions, acting on different levels of processing. Thus instead of basic emotions, only modal emotions exist who result from often found appraisal patterns (Scherer, 1994). Scherer's (2001) Component Process Theory assumes different sequential evaluation checks (SECs) in the emergence of an emotion. Those SECs include appraisals based on relevance, implications, coping potential and normative significance (Table 3).

As a response to the evaluation of relevant external or internal stimuli, a sequence of interrelated, synchronized changes of all or most organismic subsystems is initiated (Scherer, 2001). Thus, the pattern of all synchronized changes of different affective components over time constitutes an emotion (Scherer, 2001).

Table 3. Central elements of the component process model of emotion (adapted from Scherer, 2001; 2009)

Stimulus evaluation checks (SECs)	Subcategories
Relevance	Novelty, goal relevance, intrinsic pleasantness
Implications	Outcome probability, discrepancy from expectation, conduciveness, urgency
Coping potential	Agent and intention, control, power, adjustment
Normative significance	Compatibility with internal and external standards

3.2.5 Emotions and Emotion Theories in HRI

3.2.5.1 Emotion Theories Used in HRI and Affective Computing

In psychology, there is a wide array of models to describe human emotions. However, many of these models are not appropriate for formal implementation in technical systems due to their imprecise formulation. Two of the appraisal models having found widespread recognition in technical sciences is Scherer's (2001) Component Process Model (example for application in machine learning: Meuleman & Scherer, 2013) as well as the OCC-Model of Ortony, Clore und Collins (1999) (examples for implementation in Affective Computing see Calvo et al., 2015).

In Affective Computing, the cognitive approach of Ortony, Clore and Collins (OCC) has reached a certain degree of popularity due to their development of an emotion model that is computationally tractable. In a short overview, the OCC model starts with an event or object that is being appraised and reacted to emotionally as a consequence of the appraisal. The desirability of the situation determines whether positive or negative emotions are being experienced. Another computational model of emotion is Scherer's GENESE expert system (Calvo & D'Mello, 2010). The main difference between the OCC model and Scherer's (2001) Component Process Model is that the latter does not classify distinct emotions but understands distinct emotions (happiness, sadness, etc.) as classes of typical feature configurations in a multimodal dimensional space.

The theory of basic emotions also had a great influence in Affective Computing, largely due to the relative easiness of mapping a small set of universal antecedents with corresponding emotions and their associated action tendencies (Lisetti & Hudlicka, 2015). Since the fine-grained description of facial expressions based on AUs (see section 3.4) made this possible, Ekman's theory still holds a particular appeal for researchers in HRI and Affective Computing (Kappas et al., 2013; Lisetti & Hudlicka, 2015; Picard, 1997). In fact, modelling the effect of emotions via facial expressions on emotional software and hardware agents is frequently done by researchers and practitioners in HRI and Affective Computing (Reisenzein, 2015).

A comprehensive overview of emotion theories implemented by researchers and practitioners in Affective Computing and related areas can be found in Calvo et al. (2015), Krämer, Klatt, Hoffmann, and Rosenthal-von der Pütten (2013) and Marsella, Gratch, and Petta (2010).

3.2.5.2 Research on Emotional Reactions Towards Robots in HRI

The field of HRI is still young and systematic research on how users respond emotionally towards a robot remains rather scarce. The largely anecdotal evidence already suggests people are able to feel something for a robot like, for example, pity when they had to harm a robot (Bartneck & Hu, 2008) or empathetic concern for a robot that was put back in a closet (Kahn et al., 2012). One of the few systematic studies on emotional responses towards a robot was conducted by Rosenthal-von der Pütten et al. (2013) who showed participants videos of a robot being mistreated (e.g., hit on the head, punched) or being treated friendly (e.g., caressed, stroked). The authors assessed participants' physiological arousal as well as their self-reported feelings and found an increased level of physiological arousal (as assessed by skin conductance; no changes in heart rate) during the torture video. For the self-reported emotions, more positive emotions after the friendly video and more negative emotions as well as more empathetic concern was reported after the torture video and participants were also more likely to attribute feelings to the robot after the torture video. Prior interaction with the robot as well as personality traits had no influence on emotional reactions. In an extension of this study, Rosenthal-von der Pütten et al. (2014) found neural activation in limbic systems regardless of condition (whether a box, a robot or a human was harmed or treated friendly) indicating that participants did react emotionally. The authors reason that this lack of differences either indicates comparable activation patterns. Or the videos might have elicited different patterns that did not reach significance due to the less controlled stimulus material (videos) compared to pictures. The authors then compared only the human-torture videos with the robot-torture videos and found a higher neural activation in a brain region (right putamen) associated with empathy and emotional distress after watching a human being tortured compared to a robot. No further mention is made concerning the box. Regarding the self-report of emotional states, participants also rated the human-torture videos more negatively than the robot-torture videos. In general, both videos were rated more negatively than the friendly videos. Looking only at the positive affect, a difference was reported: participants felt most positively after watching a robot being treated nicely and least positive after watching a robot being tortured compared to a human (Rosenthal-von der Pütten et al., 2014). Hence, there seems to be a difference between a) physiological data, self-reports of negative affect (more negative affect for human torture) and b) self-reports of positive affect (less positive affect for robot torture). Regarding the decrease in positive feelings, which was higher for robots than for humans in the study by Rosenthal-von der Pütten et al. (2014), Unz, Schwab, and Winterhoff-Spurk (2008) also report an unexpected result concerning TV news on violence against humans and against animals (and inanimate objects): while it was assumed that participants would react more strongly to violence against humans, the opposite was the case. The authors argue that those events are appraised as more relevant and hence, more negative feelings are experienced. Furthermore, the suffering of innocent animals may have triggered commiseration (Unz et al., 2008).

While these studies already indicate profoundness in emotional reactions towards robots, some issues still remain unexplained: when do physiological data and self-reports match? Are there emotional reactions other than physiological ones or subjective feelings? And perhaps most importantly regarding application purposes: are emotional reactions visible (i.e., not depending on invasive methods like physiological measures)? Other factors also warrant further attention: due to the fMRI setting, sample size was a limiting factor ($n = 14$, Rosenthal-von der Pütten et al., 2014). Furthermore, physiological methods are inherently unspecific as they are unable to detect emotional valence or specific emotions (e.g., Arkin & Moshkina, 2015). Rosenthal-von der Pütten et al. (2014) also did not detect any differences in the reported empathy for the robot or human. However, it cannot be excluded that there were no differences at all, especially since Rosenthal-von der Pütten et al. (2013) found an effect of empathy. Considering these issues, Menne and Schwab (2018) extended the study by Rosenthal-von der Pütten et al. (2013) to include facial expressions as a visible signal of emotional experiences. Also, a between-subjects design (instead of a within-subjects design) was chosen to further eliminate influences of demand characteristics or social desirability. Analysing facial expressions of 62 participants using FACS, the authors found that particular AUs, such as lowering the brow (AU 4), commonly associated with negative emotions, were displayed while watching a video clip of a robot being tortured. Participants also reported feeling more positive after watching a robot being treated friendly and more negative after watching the torture video (Menne & Schwab, 2018). All these studies used an entertainment dinosaur, Pleo, as stimuli, whereas in another study by Menne and Lugrin (2017), Reeti, a more humanoid looking robot, was tested as a stimulus to evoke emotional reactions. The reported findings are similar to Menne and Schwab (2018). Although these studies suggest (observable) profoundness in emotional reactions towards robots, no prior study systematically compared different types of robots and the impact of their emotional expressiveness on visible emotional reactions (facial expressions) and subjective feelings (for further elaboration see section 4.1).

3.3 Empathy

The term "empathy" is difficult to grasp and recently, Cuff, Brown, Taylor and Howat (2016) found 43 different definitions. In the following, conceptualizations of empathy and similar terms are presented, empathy is set in relation to theory of mind, approaches for measuring empathy are introduced and research on empathy in HRI is highlighted.

3.3.1 Definition

Literature on empathy and definitions of empathy are inconsistent and often remain vague (Cuff et al., 2016). Whereas empathy is on the one hand viewed as a dispositional trait (Hoffman, 1982), others have emphasized empathy as a cognitive-affective state that is situation-specific (Duan & Hill, 1996). A broad definition of empathy can be found in Hoffman (2008): Empathy is "an emotional state triggered by another's emotional state or situation, in which one feels what the other feels or would normally be expected to feel in his situation" (p. 440). Hoffman (2008) presents five modes evoking empathy that can roughly be divided between conscious (verbal) and unconscious (nonverbal) processes: Mimicry (the physiological experience of feeling what another feels, 'motor mimicry'), conditioning (usually acquired through mother-infant interactions; empathic distress as conditioned response to another's display of distress), and direct association (an expression of distress in another arouses own feelings of distress) are considered to automatically and involuntarily evoke empathy, based on surface cues (Hoffman, 2008). In contrast, empathy can also be aroused through conscious processes such as perspective taking (imagining oneself in another's place) and verbally mediated associations (distress is communicated through language). This perspective is shared among many researchers: emotional empathy and cognitive empathy (cf. Davis, 1983).

Cuff et al. (2016) summarized and analyzed definitions of empathy based on 43 definitions and offer a more comprehensive conceptualization (p. 150):

Empathy is an emotional response (affective), dependent upon the interaction between trait capacities and state influences. Empathic processes are automatically elicited but are also shaped by top-down control processes. The resulting emotion is similar to one's perception (directly experienced or imagined) and understanding (cognitive empathy) of the stimulus emotion, with recognition that the source of the emotion is not one's own.

De Waal and Preston (2017) point out that emotional and cognitive empathy "remain interconnected in evolution, across species and at the level of neural mechanisms" (p. 498). They report a perception-action mechanism (PAM; for a review see de Waal, 2008; de Waal & Preston, 2017), the core for empathy, that automatically activates neural representations of states, causing a match in emotional states between subject and object (de Waal, 2008).

The terms 'emotional contagion' and 'sympathy' are also often mentioned in the context of empathy. In contrast to emotional contagion and empathy, sympathy can be viewed as feeling regret for the other without necessarily experiencing a corresponding emotional state (Preston & de Waal, 2002). Differences between self-other-awareness have also been discussed (e.g., Hoffman, 1975; de Waal, 1996).

Emotional contagion is defined as "the tendency to 'catch' (experience/express) another person's emotions" (Hatfield, Cacioppo, & Rapson, 1992, p. 153). Hsee, Hatfield, Carslon, and Chemtob (1999) for example could show that facial expressions of happiness and sadness, expressed by a confederate, were mirrored by participants. They displayed more facial expressions of happiness and less sadness when the confederate expressed happiness and vice versa. This was further matched by self-reports of participants' emotional states.

Scherer (1998) introduces another concept: he calls emotions that arise from the perception of the emotions of others "commotions". Hence, Scherer (1998) proposes that empathy, besides induction (appraisal processes) and emotional contagion, is a mechanism that triggers commotions. Disentangling the terms emotional contagion from empathy proofs to be relatively difficult as definitions are inconsistent. Some authors view emotion contagion as accounted for by processes of empathy (e.g., Kelly & Barsade, 2001) while others consider empathy and emotional contagion as equivalent or view emotion contagion as a process resulting in empathy (e.g., Eisenberg & Miller, 1987; Levenson, 1996; Scherer,

1998; Vaughan & Lanzetta, 1981). Considering these issues, this dissertation uses the terms 'emotional contagion' and 'empathy' synonymously.

3.3.2 Empathy and Theory of Mind

The term "Theory of Mind" was first used by Premack and Woodruff (1978) and describes the ability to attribute mental states (e.g., knowledges, desires, thoughts and emotions) to oneself and another being (Pedersen, 2018; see also Schneider & Lindenberger, 2018). Developmental studies have shown that even 6 month old infants have an intuitive knowledge that human actions are oriented towards specific goals (intentionality). Studies using the false belief paradigm investigate a person's ability to understand mental states of others. Developmental research shows that children between four and five mature to understand that others can have false beliefs and act upon them; and they are able to correctly predict actions from those false beliefs (e.g., Wimmer & Perner, 1983). Regarding emotions, newborns cry in response to distress of other infants and eight week old infants mirror positive emotions in their facial expressions, showing that the ability for empathy exists from early on (Elsner & Pauen, 2012). As children grow older, their language skills further develop. Thus, between the ages of three and seven, children become more competent in talking about negative emotions and setting them in relation to causes (e.g., sadness associated with loss) (Hughes & Dunn, 2002). The maturing process eventually leads to an individual's ability for complex evaluation, reflection and analysis of emotions. For an overview of emotions, media and childhood development see Nieding and Ohler (2018).

Whether empathy has to be distinguished from terms like 'theory of mind' (Premack & Woodruff, 1978; Whiten, 1991; see Call & Tomasello, 2008, for a review), 'mentalizing' (Frith & Frith, 2003) or 'cognitive perspective taking' (Saxe, 2006) is still discussed (Batson, 2009).

3.3.3 Gender and Individual Differences

Women are considered to be more empathic (e.g., Cheng, Tzeng, Decety, & Hsieh, 2006; Eisenberg & Fabes, 1990; Eisenberg & Lennon, 1983). However, findings are inconsistent and seem to depend on the methods and definitions used for empathy (e.g., Eisenberg & Fabes, 1990, for gender differences in self-report studies). Systematic research on gender differences in HRI is still at the beginning and largely lacking robust findings. One study on emotional reactions

towards robots did not find an effect of gender on emotional state or physiological arousal (Rosenthal-von der Pütten et al., 2013). However, gender differences
have been found for negative evaluation of a humanoid robot and anxiety to interact with it (de Graaf & Allouch, 2013), and in anthropomorphization
(Schermerhorn, Scheutz, & Crowell, 2008). Women perceived a rabbit-shaped
robot as more positive than men (Eimler, Krämer, & von der Pütten, 2011), were
more likely to follow the advice of a catlike robot (Vossen, Ham, & Midden, 2009)
and showed greater willingness to form a relationship with pet-like robots (Fujita, 2004; Turkle, 2011). However, findings are inconsistent (e.g., de Graaf and
Allouch, 2017, Rosenthal-von der Pütten et al., 2013).

Research in psychology has shown that people's individual dispositions such
as empathy trait (Davis, 1983), affiliative tendency (Mehrabian, 1976) and loneliness (Russell, Peplau & Cutrona, 1980) affect emotional responses on a broad
scale (e.g., Davis, 1983). Archer, Diaz-Loving, Gollwitzer, Davis, & Foushee
(1981) for example, could show that a dispositional empathic tendency was associated with empathic concern and personal distress. The need to belong
(Baumeister & Leary, 1995), humans' desire to bond with others, is closely connected to affilative tendency and loneliness, and has been shown to be linked with
social effects of virtual agents (e.g., Krämer, Lucas, Schmitt, & Gratch, 2018) or
users' willingness to bond with artificial beings (e.g., Krämer, Eimler, von der
Pütten & Payr, 2011). Rosenthal-von der Pütten et al. (2013) also included these
factors but did not find a mediating effect on emotional reactions towards a robot. However, this might have been due to the homogeneous sample (Rosenthal-
von der Pütten et al., 2013). Hence, the effect of individual dispositions on emotional reactions towards robots will be further explored in this doctoral dissertation.

3.3.4 Measuring Empathy

Empathy is usually defined as an emotional state (e.g., Hoffman, 2008) and thus
can be found in most handbooks on emotion. Even though there are specific
methods for measuring empathy (the most widely used self-report measure is the
Interpersonal Reactivity Index, IRI, Davis, 1983), the term is still vague and many
studies generally refer to a broader definition to encompass a wider range of emotional experiences (Cuff et al., 2016; in HRI, e.g., Rosenthal-von der Pütten et al.,
2013). Furthermore, observational and physiological methods for measuring empathy do not differ significantly from those described in section 3.5.2.2. For example, measuring heart rate and skin conductance and correlating the results

with self-report measurements (e.g., Levenson & Ruef, 1992). Or using Facial EMG to measure motor mimicry (Neumann, Chan, Boyle, Wan, & Westbury, 2015). Similar to the claim to use a multi-method approach in affective HRI (Arkin & Moshkina, 2015) and for measuring emotions (e.g., Scherer, 2005), Neumann et al. (2015) also advocate a combination of measures "to provide a comprehensive approach to empathy assessment" (p. 285). They further elaborate: "Such a battery may best comprise a broad self-report measure of empathy administered in conjunction with appropriate behavioral tests or physiological measures of empathy in specific situations, such as in emotional contagion, motor mimicry, or empathy for pain" (p. 285).

3.3.5 Empathy in HRI

In general, as with studies on emotional reactions towards robots, systematic studies on empathic reactions are still at the beginning in social robotics research. Studies can roughly be classified into two categories: either the focus is on creating an 'empathic' robot and a (brief) evaluation (e.g., Hegel, Spexard, Wrede, Horstmann, & Vogt, 2006; Leite, Castellano, Pereira, Martinho, & Paiva, 2014; Riek, Paul, & Robinson, 2010; Zeng, Pantic, Roisman, & Huang, 2009) or how robots evoke empathy (and emotional reactions in a broader sense) in humans and evaluating the effects (e.g., Gonsior et al., 2011; Gonsior, Sosnowski, Buss, Wollherr, & Kuhnlenz, 2012; Jo, Han, Chung, & Lee, 2013; Kwak, Kim, Kim, Shin, & Cho, 2013; Menne & Lugrin, 2017; Menne & Schwab, 2018; Riek, Rabinowitch, Chakrabarti, & Robinson, 2009; Rosenthal-von der Pütten et al., 2013; Rosenthal-von der Pütten et al., 2014).

Concerning the creation of empathic robots, effects on, for instance, believability, anthropomorphism, friendship, and perceived intelligence (Bartneck, Kulić, Croft, & Zoghbi, 2009) or intention to use the agent (e.g., Heerink, Kröse, Evers, & Wielinga, 2010) have been measured as an indication of perceived empathy.

For investigations on a robot's potential to evoke empathy, some studies (Menne & Lugrin, 2017; Menne & Schwab, 2018; Rosenthal-von der Pütten et al., 2013; Rosenthal-von der Pütten et al., 2014) have already been described in section 3.2.5.2 to reflect common conceptualizations of empathy as part of an emotional state (see section 3.3.1).

The following studies on robots evoking empathy do not use a similar paradigm as those mentioned before and are thus briefly introduced in this section. Gonsier et al. (2012) presented participants a robotic head able to mirror facial expressions of the user. After a small talk interaction with the robotic head, participants could help the robotic head in classifying objects in a picture. Participants showed higher helpfulness towards the robot that showed emotional behavior. Furthermore, they rated it to be more humanlike and attentive than those in the control group (no emotions expressed). In another study by Gonsier et al. (2011) the same robotic head was used to play a game where the robotic head tried to guess a person the users were thinking of. Participants reported to feel more empathy towards the robot when it expressed emotions and mirrored participants' facial expressions than when the robot did not express emotions. Hayes, Ullman, Alexander, Bank, & Scassellati (2014) reported that in a counting game, participants were more likely to help the Keepon robot (a small yellow toy robot, looking like a small 'snowman', cf. Kozima, Michalowski, & Nakagawa, 2009) speaking in an expressive voice about its own or other-related distress than when the robot did not express any distress. In a study inspired by Milgram (1963), Kwak et al. (2013) let children administer electric shocks to a 'learner', an egg-shaped robot (robot Mung, Kim, Kwak, Hyun, Kim, Kwak, 2009) that is able to show 'bruises' by different coloring of its elastic skin. That way, the robot was able to express a negative emotional state. The level of agency was manipulated by either showing participants a picture of a remote user controlling the robot (robot as a mediator) or the robot being autonomous (robot is situated). Results show that the children empathized more with the robot acting as a mediator. Self-reports were used to assess emotional states. In a web-based study, Riek and colleagues (2009) studied the impact of human-likeness of a robot and the effect on empathic responses. Using fictional films containing scenes about robots being mistreated, participants rated on a single item how sorry they felt for the protagonist after each video. Furthermore, they were asked to indicate which of the robots they would save in an earthquake and completed a trait empathy questionnaire. Results show that participants were more empathetic toward human-like robots and less toward mechanical-looking robots (e.g., Roomba). However, the content of each clip was different for the protagonists, which produced a potential confound: Causal inferences attributing participants' ratings to the human-likeness of a robot cannot be drawn since the clip content was not the same for every protagonist in the video.

Summarizing research on robots evoking empathy, there is a wide variety of different methods for assessing empathic reactions and no consensus has been reached (Paiva, Leite, Boukricha, & Wachsmuth, 2017). For example, Riek et al. (2009) used a single item to measure situational empathy. Kwak et al. (2013) calculated difference scores between own and other-oriented empathy. As with research on emotional reactions in general, studies on empathic reactions largely lack a rigorous experimental approach to disentangle causal effects (see Eyssel,

2017, for an overview). Furthermore, a multi-method approach to measuring empathy has been scarce. Even though there are exceptions (e.g., Rosenthal-von der Pütten et al., 2014, who used fMRI as well as self-reports to measure empathic reactions), those are often limited, for instance, in sample size, due to methodological constraints and thus lack in statistical power and restraints in correctly using inferential statistics which is a problem that also concerns social robotics research in general (see Eyssel, 2017, for an overview). Self-report measurements for assessing empathic responses are mainly constructed specifically for the study's needs (e.g., Gonsior, et al., 2011; Riek et al., 2009). For example, Gonsior et al. (2011) studied effects of mirroring facial expressions of a robot, but instead of measuring participants' mirroring of the robot's facial expressions, self-reports of scenario-specific statements intended to assess induced empathy were used.

3.4 Facial Expressions

93% of meaning comes from nonverbal channels. This figure, originally published by Mehrabian (1981), is frequently cited in a wide variety of fields such as HRI (e.g., Pantic & Rothkrantz, 2000; Lee, Park, Jo, & Chung 2007; Park, Lee, & Chung, 2015). However, as Burgoon (2013) among others (e.g. Lapakko, 1997) point out, the design of the original study does not allow for this generalization. The dominance of the nonverbal channel has been discussed among researchers (e.g., Scherer, Scherer, Hall, & Rosenthal, 1977) and it could be shown that nonverbal behavior is especially important for social relationships and impression formation: Adults trust nonverbal cues rather than verbal cues in general (Burgoon, 2013). Furthermore, when evaluating the emotional state, nonverbal information is preferred (Burgoon, 2013).

Facial expressions are by far the most frequently studied nonverbal communication channel with 95% of studies on emotions in humans having used facial expression stimuli (de Gelder, 2009). In this section, the relevance of the face and facial expression for interaction is described. The Facial Action Coding System (FACS) as the most widely and most frequently used method for analyzing facial expressions is presented. Since FACS coding is time intensive, alternatives like automatic recognition of facial expressions are briefly discussed. Finally, research using facial expressions in human-human interaction as well as in human-robot interaction is illustrated.

3.4.1 Importance of the Face and Facial Expressions

Face-to-face communication is inherently natural and provides an efficient way to exchange information without changing habits (Jaeckel, Campbell, & Melhuish, 2008). As the face plays such a central role for human interaction, it would be uneconomical to ignore this rich communication channel (Bartneck & Lyons, 2009) and unnecessarily invent alternative ways for communication. This is especially important considering human-robot interaction. With robots entering our domestic lives, unspecialized lay people come into contact with robots. However, most people know how to interact with other humans and even animals using verbal and nonverbal communication. It would be unwise not to capitalize on this existing ability and hence, current research aims to equip robots with the ability to assess affective displays of the human communication partner and respond appropriately with familiar human social cues.

In the course of evolution, the communication of thoughts, information and emotions between humans found many ways to be expressed, starting from nonverbal expression to spoken language. Housing the majority of our sensory organs (eyes, nose, mouth, ears), the face plays a main role in the communication process both for nonverbal and verbal communication and is considered to be the most important mode of nonverbal communication (Ambady & Rosenthal, 1992; Ekman & Friesen, 1969; Ekman & Rosenberg, 2005). This predominant role is rooted deeply in our language as we speak of "face-to-face interaction", "interface", or even "facebook". Even though nonverbal communication is a multimodal process, the visual sense is the most important channel. Hence, more research focuses on the face and facial expressions than on any other nonverbal channel (Kappas et al., 2013). Furthermore, humans have a functional sensitivity for faces and facial expressions since faces present an extraordinarily potent emotional and social stimuli (Haxby, Hoffman, & Gobbini, 2000; Öhman, 2002). Research has even identified brain regions specialized for face perception (e.g., Kanwhisher, McDermott, & Chun, 1997; Cohen Kadosh & Johnson, 2007).

3.4.2 Facial Action Coding System

The Facial Action Coding System (FACS) by Ekman & Friesen (1978) (revised version: Ekman, Friesen, & Hager, 2002) is the most comprehensive and most widely used method for analyzing facial expressions (see Ekman & Rosenberg, 2005, for an overview). It is used in a wide variety of fields, including computer vision (e.g., Bartlett, Hager, Ekman, & Sejnowski, 1999; Ko, 2018; Lien, Kanade,

Cohn, & Li, 2000), robotics (e.g., Menne & Schwab, 2018) and social studies of emotion (e.g., Ekman & Rosenberg, 2005).

FACS is an objective and standardized method for measuring visual appearance changes in the face (Ekman et al., 2002). The smallest units of muscular activity visible in the human face are Action Units (AUs). A single Action Unit (AU) or combinations of several AUs describe every visible movement in the face. The coding and interpretation procedure usually runs in two steps. First, external signs of facial behavior, such as lowering the brows, are coded and assigned their specific AU (brow lowerer: AU 4). Second, if the empirical evidence is sufficient, those AUs are interpreted as indicators of psychological processes. One of the major strengths of FACS is that it refrains from imposing meaning categories but allows an objective description of appearance changes in the face.

3.4.3 Automatic Recognition of Facial Expressions

While manual coding of facial expressions is a taxing and time consuming task with coders requiring expert training in FACS, automatic recognition of facial expressions seems a promising alternative. Indeed, with advances in artificial intelligence (Human-Computer Interaction: e.g., Dornaika & Raducanu (2007) or entertainment: e.g., Zhan, Li, Ogunbona, & Safaei, 2008), interest in automatic facial emotion recognition has increased rapidly (Ko, 2018). However, most studies focus mainly on the recognition of a small range of basic emotions (Martinez & Valstar, 2016; Fasel & Luettin, 2003) or concentrate on acted rather than spontaneous facial expressions (Kawulok, Celebi, & Smolka, 2016). Furthermore, the automatic analysis of AUs still has to work on a set of problems, such as the reliable automatic detection of AUs (Valstar, Mehu, Jiang, Pantic, & Scherer, 2012), the automatic intensity coding of AUs (Kaltwang, Todorovic, & Pantic, 2015) or the automatic identification of temporal segments of AUs (Jiang, Martinez, & Pantic, 2014).

3.4.4 Facial Expressions and Emotion

In a natural interaction, the communication partner is usually able to infer the affective state of his counterpart based on external observable cues. Cues that are unobtrusively observable are facial expressions. Facial expressions can be seen as direct and naturally preeminent ways for communicating emotions (Matsumoto, Keltner, Shiota, O'Sullivan, & Frank, 2008; Russell & Fernández-Dols, 1997; see also Mauss & Robinson, 2009, for a review). First, research linking (spontaneous) facial expressions to emotions in psychology (human-human interaction) is reviewed, and then research in HRI involving the analysis of facial expressions and emotion is illustrated.

3.4.4.1 Research on Facial Expressions in Human-Human Interaction

Matsumoto et al. (2008) summarize findings from literature about facial expressions of emotion (p. 227):

Table 4. Characteristics of facial expressions of emotion (Matsumoto et al., 2008)

1	Some facial expressions are universal, reliable markers of discrete emotions when emotions are aroused and there is no reason to modify or manage the expressions.
2	Discrete facial expressions generally correspond to discrete underlying subjective experiences.
3	Discrete facial expressions are part of a coherent package of emotion responses that includes appraisals, physiological reactions, other nonverbal behaviors, and subsequent actions; they are also reliable signs of individual differences and of mental and physical health.
4	Discrete facial expressions are judged reliably in different cultures.
5	Discrete facial expressions serve many interpersonal and social regulatory functions.

Most important to this dissertation is evidence from associations between spontaneous facial expressions and subjective experiences. Research suggests there is a reliable connection between those (e.g. Bonanno & Keltner, 1997, 2004; Ekman, Friesen, & Ancoli, 1980; Ekman, Davidson, & Friesen, 1990; Harris & Alvarado, 2005; Mauss, Levenson, McCarter, Wilhelm, & Gross, 2005; Rosenberg & Ekman, 1994; Ruch, 1994; Ruch, 1995). Only some examples of emotional facial expression research are given (see Ekman & Rosenberg, 2005, for an overview).

Results from studies using FACS to code facial expressions as well as self-reports of emotional experience have found positive associations. A study by Unz et al. (2008) showed that violence in TV news triggers negative facial expressions (primarily AU 14) as well as negative feelings. They also mention that participants reacted more strongly to violence against animals than to violence against inanimate objects or humans. Ekman et al. (1980) reported that participants who displayed activity of the zygomaticus major in response to pleasant films reported feeling happier. Likewise, those who displayed negative facial action units when watching unpleasant films reported feeling more fear, disgust, surprise, pain and arousal than those who did not show emotion-specific facial expressions. The number of relevant AUs shown was also positively correlated with self-reported intensity of positive and negative feelings. Furthermore, facial expressions like wrinkling of the muscles around the eye and lifting the corners of the lips ("Duchenne smile") are associated with positive emotions (Ekman et al., 1990; Frank, Ekman, & Friesen, 1993; Hess, Banse, & Kappas, 1995; Keltner & Bonanno, 1997). Eyebrow lowering has been linked to negative emotions (Kring & Sloan, 2007).

Even though some of these studies have shown that facial expressions are associated with specific emotions, Mauss and Robinson (2009) point out that facial expressions seem to be particularly sensitive to the valence of a person's emotional state (see also Russell, 1994). Indeed, many findings on emotional facial expressions appear to be along this line of thought (although cf. Ekman et al., 2013, for a more detailed discussion). Mauss et al. (2005) could show high correlations between facial behavior and the valence of emotional experience for films inducing either amusement or sadness. Examples not using FACS, but users evaluating facial expressions (judgement approach, cf. Robinson, 2005) come from studies using the slide-viewing paradigm (Buck, 1978; Buck, Miller & Caul, 1974. Participants viewed emotionally evocative colored slides while being recorded with a hidden camera. Then, other participants are asked to watch the videotaped facial expressions of those prior participants and identify the type of slide that has been watched as well as evaluate the pleasantness of the prior participant's emotional reaction. It could be shown that the latter participants could accurately identify the slide category that has been viewed by other participants before. Furthermore, positive correlations between prior participants' self-reported pleasantness ratings of the slides and later participants' pleasantness ratings have been found, indicating that observers are able to identify the pleasantness of emotional reactions from a sender's face alone.

Studies using facial electromyography (EMG) (measuring electrical potential from facial muscles by placing electrodes on the face) have found correlations between muscle activity and the valence of emotional experiences. As such, the corrugator supercilii (associated with lowering of the eyebrows) is linked to the unpleasantness of an affective stimuli whereas the zygomaticus major (associated with raising of the lip corners) is positively correlated with the pleasantness of an affective stimuli (Dimberg, 1982; Dimberg & Thunberg, 1998; see Bradley & Lang, 2000; Lang, Greenwald, Bradley, & Hamm, 1993; Larsen, Norris, & Cacioppo, 2003, for reviews). Sato and Yoshikawa (2007) could show that not only (mostly invisible) facial muscle activity corresponding to affective stimuli occurred but that these muscle activities produced visible appearance changes. By using FACS, the authors found that brow lowering was linked to dynamic presentations of angry expressions, whereas raising of the lip corners could be observed for happy expressions.

An abundance of research has shown that facial expressions appear to be sensitive to emotional experience and researchers in technical fields rely on these findings when investigating emotional experiences.

However, with still no comprehensive emotion theory, caution is advised when interpreting facial expressions as emotional states since research has shown that facial expressions are not only associated with emotional experiences but serve many functions (cf. Barrett, Mesquita, & Gendron, 2011). Furthermore, some factors (sex, culture, expressiveness, audience) moderate the link between emotional states and facial expressions (Mauss & Robinson, 2009).

3.4.4.2 Research on Facial Expressions in HRI

Although an increasing body of work in HRI focuses on facial expressions, social robotic research is almost solely concerned with either creating facial expressions on robots (mostly by using FACS) (e.g., Bartneck, 2002; Becker-Aasano, Ishiguro, 2011; Breazeal, 2003; Hegel, Eyssel, Wrede, 2010; Sosnowski, Bittermann, Kuhnlenz, & Buss, 2006; Wu, Butko, Ruvulo, Bartlett, & Movellan, 2009) or recognizing facial expressions posed by robots (e.g., Costa, Soares, & Santos, 2013; Endo et al., 2008; Mirnig et al., 2015; Takahashi, & Hatakeyama, 2008).

There is mainly anecdotal evidence of facial expressions of emotion towards robots (e.g. Bartneck & Hu, 2008; Breazeal, 2002b; Heerink, Kröse, Evers, & Wielinga, 2009). Breazeal (2002b) for example, reported subjects reacted empathetic towards the saddened face of a robot. One participant would "look to the experimenter with an anguished expression on her face, claiming to feel 'terrible' or 'guilty'" (p. 899). Bartneck and Hu (2008) report participants "showed compassion for the robot" (p. 420) or "felt bad" (p. 426) and reactions similar to those in the Milgram paradigm (Milgram, 1974) ("giggled" or "laughed" to relieve the pressure, Bartneck & Hu, 2008, p. 428) when participants had to harm a robot.

A study with a focus on particular facial muscles found a deactivation of corrugator supercilii (involved in AU 4, 'brow lowerer') and an activation of zygomaticus major (associated with AU 12, 'lip corner puller') in response to happy robotic faces (Riether, 2013). In contrast, viewing sad robotic faces resulted in an increase of corrugator supercilii activation and a decrease of zygomaticus major activation (Riether, 2013). Those results are similar to Dimberg (1982) and Dimberg and Thunberg (1998), suggesting people may react the same towards human and robotic emotional faces. Likewise, Cañamero (2002) observed participant's facial expressions matched those of the robot Feelix. Imitating facial expressions is an unconscious reaction involved in the process of sympathizing with someone else (Davis, 1983). Hence, studying facial expressions presents an opportunity for analyzing the profoundness of emotional reactions towards robots. Systematic research on FACS-coded facial expressions shown towards (different types of) robots has been conducted by Menne and Schwab (2018) as well as Menne and Lugrin (2017) (see section 3.2.5.2 for further details).

3.5 Measurement of Emotions

Shiota and Kalat (2012) point out that "scientific progress almost always depends on improved measurements" (p. 356) and that "measurement is an important and contentious issue in any area of science" (p. 356). They advocate using multiple measures of emotion (Shiota and Kalat, 2012) to compensate disadvantages of different measurements and obtain convergent validity. In the following section, the Media Equation, an approach popular in technical fields, will be introduced briefly, focusing on methodological issues with measuring psychological reactions towards artificial entities. Following this, a brief overview of methods used in emotion research in psychology and HRI will be given, focusing mainly on those relevant for this dissertation. Furthermore, advantages and disadvantages using self-reports, observational and physiological methods will be discussed and evaluated in a concluding remark.

3.5.1 Media Equation and Matters of Measurement

Certainly the most popular statement of the Media Equation approach, adapted from the title of a book by Reeves and Nass (1996) is that people mindlessly treat computers, TV, and new media like real people. Indeed, humans are able to attribute life and affect to inanimate objects (Melson, Kahn, Beck, & Friedman, 2009; Reeves & Nass, 1996). Robots share more features similar to humans and other beings than personal computers (the original research object by Reeves and Nass, 1996) and virtual characters, enabling them to further increase impressions of 'being alive'. For example, next to their physical presence, robots are also able to show emotions, for instance, by using facial expressions (e.g., Wu et al., 2009) (note that until now, robots are only able to convey the impression of an emotion). The implications of these phenomena are reflected in concerns for the well-being of robots (e.g., Whitby, 2008) and even establishing a research program on robot ethics (cf. Lin, Abney, & Bekey, 2014).

Next to robots taking the Media Equation approach to the next level, issues for measuring the mostly automatic and unconscious reactions to media entities (Reeves & Nass, 1996) are discussed. Reeves and Nass (1996) argue that those reactions are fundamentally rooted in evolved human nature and thus emphasize, that "attempts to verify the media equation can't rely solely on talking to people, (…) or asking them questions on a survey" (p. 7). Hence, the following sections will focus on different methods for measuring emotional reactions.

3.5.2 Methods for Assessing Emotions

No single "gold-standard method" (Scherer, 2005, p. 709) appears to exist for measuring emotions. However, two aspects are important: first, to measure at the appropriate point in time since most of the time, emotions only last less than some seconds (Ekman, 2009). And second, to choose an appropriate measure from the broad range of parameters (e.g., subjective feelings, facial expressions, heart rate, etc.). Different methods have been established for measuring emotions: self-reports, physiological methods and behavior analysis. Since this dissertation focuses on the measurement of emotions via self-reports and observational methods (the analysis of facial expressions), these will be illustrated here, while physiological methods will only be briefly addressed (but see e.g., Lewis, Haviland-Jones, & Feldman-Barrett, 2008, for a more detailed view on physiological approaches).

3.5.2.1 Subjective Methods: Self-Reports

The subjective feeling component can only be assessed through self-reports (Shiota & Kalat, 2012). Compared to, for example psycho-physiological measures, self-reports are far more economical in terms of experimental effort, skills of experimenter, skills for analysis, and cost efficiency which is why most researchers use self-reports to capture emotional states (Weidman, Steckler, & Tracy, 2017).

Research has shown that self-reports of current experiences is likely to be more valid than self-reports on past or future experiences of emotion (Robinson & Clore, 2002). However, several factors also impact the validity of self-reports of one's current emotional experience: First, there is reason to believe that individuals high in social desirability will not (impression management) or cannot (self-deception) validly report their (negative) emotional state (Paulhus & Reid,

1991; Paulhus & John, 1998; Welte & Russell, 1993). Second, not only self-reports from those high in social desirability may suffer from invalidity but also vulnerable populations like babies, individuals suffering from brain injury or otherwise mentally incapacitated individuals may not be able to give valid self-reports. Third, even though one may experience emotions, but may not be able to consciously express them via self-reports (Lane, Ahern, Schwartz, & Kaszniak, 1997). Fourth, awareness for one's own emotions may not always be given (see e.g., Lesser, 1981, for a review on the alexithymia concept and Baron-Cohen, 2000, for a review on autism). These issues present a major drawback in the use of self-reports to reliably and validly capture individuals' emotional experience.

Next to these limitations in individuals' self-reports, there are also methodological pitfalls in the assessment of emotional experiences, especially concerning the measurement of distinct emotions. Whereas self-report measurements of dimensional aspects of emotions are already established, such as the Positive and Negative Affect Schedule (PANAS; Watson, Clark, & Tellegen, 1988) which is widely and frequently used (Weidman et al., 2017), no comprehensive means of measuring distinct emotions exist (Weidman et al., 2017). In a recent review of self-reports of distinct emotions, Weidman et al. (2017) came to the conclusion that most researchers tend to use "short, impromptu scales" (p. 290) with unknown psychometric criteria. The authors suggest to draw on existing research and theory for scale construction, especially for the assessment of distinct emotions. Examples for psychometric questionnaires on distinct emotions are the Differential Emotion Scale (DES) by Izard, Dougherty, Bloxom and Kotsch (1974) and its variations (e.g., Modified Differential Affect Scale [MDAS] by Renaud and Unz, 2006) as well as the Multidimensional Anger Inventory (Siegel, 1986) and the Buss-Durkee Hostility Inventory (Buss & Durkee, 1957) for specific emotions.

While the assessment of discrete emotional states via self-reports may still be challenging, several authors favor the dimensional approach as it seems to capture most of the variance of emotional experience (Mauss & Robinson, 2009; Russell & Barrett, 1999; Watson, 2000). However, assessing distinct emotions can prove to be a beneficial approach since "contemporary affective science has seen a surge of interest in distinct, momentary emotional states" (Weidman et al., 2017, p. 267).

3.5.2.2 Objective Methods: Observational and Physiological Methods

While self-reports, as described in section 3.5.2.1, suffer from certain limitations, such as social desirability, self-deception, inability to verbalize emotional feelings, unawareness of emotional feelings, difficulties in reconstructing emotional feelings post-hoc, greatly limiting reliability and validity (e.g., Austin, Deary, Gibson, McGregor, & Dent, 1998; Fan et al., 2006; Wilcox, 2011), objective methods do not rely on participants' capability and willingness to accurately describe emotional experiences. From the point of view of the component approach to emotions, different components are being measured by subjective and objective methods: self-reports measure subjective feelings whereas objective methods capture observational (i.e. behavioral) and physiological aspects of emotions.

There is a multitude of observational measures to capture different aspects of behavior as indicator of emotions (e.g. vocal characteristics, whole-body behavior, and facial expressions). The analysis of facial expressions via FACS is most relevant to this dissertation and is described in an own section (section 3.4). Here, the focus lies on only a brief overview of advantages and disadvantages of observational and physiological methods for assessing emotions. For a more detailed review see Mauss and Robinson (2009).

Observational methods for assessing emotions include the analysis of facial expressions (most widely and frequently used method: FACS by Ekman et al., 2002), vocal characteristics (for example the voice amplitude and pitch, see e.g., Kappas, Hess, & Scherer, 1991) and body behavior (for example body posture, see e.g., Tracy & Robins, 2004).

Physiological methods involve (peripheral) physiological data (skin conductance, blood pressure) and neuropsychological methods such as EEG, EMG, fMRI or CT (Larsen, Berntson, Poehlmann, Ito, & Cacioppo, 2008). The aim is to detect physiological correlates of psychological processes. For example, correlations between heart rate and increased arousal have been studied (in media research: Ravaja, 2004; psychophysiological correlates in HRI: e.g., Kulic & Croft, 2007). These methods allow measuring objective physiological processes but are limited in their specificity: only the overall arousal level can be measured making the use of supplementary measures such as self-reports necessary for cross-validation and additional information.

Since facial Electromyography (EMG) is a physiological method for measuring muscular activity in the face and is often used in behavioral studies on facial expressions, it will be briefly described here. Electrical potential from facial muscles indicative of emotion is measured by placing electrodes on the face. That way, activation in the muscles of corrugator supercilii (associated with brow lowering) and zygomaticus major (associated with pulling the lip corners) can be measured even though they may not be visible in the face. Research has shown that activation of the zygomaticus major increases linearly with the pleasantness of affective stimuli while activation of the corrugator supercilii decreases linearly with the pleasantness of affective stimuli (see Bradley & Lang, 2000; Larsen et al.,

2003, for reviews). Facial EMG is thus able to provide information about emotional valence but only to a very limited extent about specific emotions. Observational and physiological methods are considered more objective since they measure behavioral and physiological aspects that cannot be easily manipulated by participants (particularly neurophysiological factors) and are therefore free from self-report bias, making them suitable for studying sensitive matters (Ravaja, 2004). They rely on the assumption that human psychological aspects are "embodied and embedded phenomena" (Cacioppo, Tassinary, & Berntson, 2007, p. 14). As such, when using physiological methods, an array of physiological sensors are usually applied to a participant, that are obtrusive in nature and thus interfere with natural body movement, are uncomfortable, especially when having to be worn for a longer time duration, etc. (see Cacioppo et al., 2007, for a review). Observational methods such as FACS avoid the obtrusiveness induced by physiological methods and thus make them more suitable for studying, for instance, emotional processes or natural interactions without necessarily alerting the participant to its measure (reactivity). However, observational methods usually require extensive training and high workload in coding (especially regarding FACS) and analyzing (both physiological and observational methods).

3.5.2.3 Methods for Measuring Emotions in HRI

Social robotics is still in its very early stages and methods specifically developed for HRI for measuring psychological processes in general and emotional reactions in particular are far from being established. Indeed, it is reasonable to argue that HRI should not be separated from the century-long psychological research on emotion (see also Calvo & D'Mello, 2010). Thus, key paradigms from psychology have already been adapted in HRI (e.g., Scassellati, 2006). Self-report measurements largely prevail in HRI (Arkin & Moshkina, 2015; Bethel & Murphy, 2010a;). The most commonly used self-report for measuring emotional reactions in HRI is the Positive and Negative Affect Schedule (PANAS) by Watson et al. (1988) (e.g., Rosenthal-von der Pütten et al., 2013; 2014; see also Arkin & Moshkina, 2015). However, the use of psychometric tests is by no means self-evident as the wide array of self-constructed questionnaires in HRI, applied in an ad hoc manner, shows (Bartneck et al., 2009). One of the main reasons for this is doubtless the fact that HRI is a young and highly interdisciplinary field involving many challenges (e.g., Bethel & Murphy, 2010a; Eyssel, 2017). Likewise, methods other than self-reports have been applied in a similar manner and for HRI and social robotics, Eyssel (2017) advocates for "unified standards regarding measurement objectivity, reliability, and validity, as well as research and data analysis practices" (p. 365) to "facilitate cross-disciplinary review processes, quality control, resulting in true advancement of science beyond disciplinary boundaries" (p. 365).

3.5.2.4 Conclusion: How are Emotions Best Measured?

Scherer (2005) stated: "there is no single gold-standard method" for measuring emotions. Indeed, difficulties in measuring emotions is the central issue of emotion research in psychology and thus also in HRI. It largely depends on the emotion conceptualization used. When viewing emotions as consisting of different components, only the measurement of all components included "can provide a comprehensive measure of an emotion" (Scherer, 2005, p. 709). As such, for the subjective feeling component, a self-report measurement (such as the PANAS for instance) can be applied. Regarding behavioral and physiological components, corresponding methods such as analyzing facial behavior, measuring skin conductance level etc., can be used. However, "such a comprehensive measurement of emotion has never been performed" (Scherer, 2005, p. 709). Indeed, self-report measurements of emotions are used almost exclusively in psychology as well as in HRI (Bethel & Murphy, 2010a; Arkin & Moshkina, 2015). Hence, if emotions are seen as a multi-level phenomenon (which is agreed upon by most researchers) – then a multi-method approach is required to adequately capture the entirety of an emotional episode. Practically speaking, Bethel and Murphy (2010a) as well as Arkin and Moshkina (2015) advocate to use "more than a single method of evaluation to obtain comprehensive understanding and convergent validity in assessments of affective HRI" (Arkin & Moshkina, 2015, p. 491).

3.5.3 Online Research Versus Laboratory Research

Another important issue to consider when conducting research is the question whether to run an experimental study in a laboratory, i.e. 'live' or an online experimental study (i.e. web-based study). In the following, a short overview on advantages and disadvantages of online research versus laboratory research will be given.

Conducting research online is increasingly popular among researchers in psychology and HRI, which is reflected in the massive use of 'paid online workers' alone, such as in Amazon Mechanical Turk (see e.g., Buhrmester, Kwang, & Gosling, 2011, for an overview and Paolacci & Chandler, 2014, for an evaluation).

Online research undoubtedly has many advantages over common laboratory research: they provide a quick and easy way for data collection, allow access to more diverse and larger samples and low cost (Birnbaum, 2004; Reips, 2002). In contrast, online research suffers from lack of control, increased participant dropout and repeated participation (Birnbaum, 2004; Reips, 2002). However, one of the main drawbacks of online research is the concern of poorer data quality (e.g., Chandler, Mueller, & Paolacci, 2014). Furthermore, results from laboratory research and online research may not be comparable (Birnbaum, 2004; Dandurand, Shultz, & Onishi, 2008; Reips, 2002; but see Dandurand et al., 2008; Germine et al., 2012; Lewis, Watson, & White, 2009, for different perspectives).

A part of this dissertation was thus dedicated to test whether results from an online experimental research in HRI are equivalent to a laboratory experimental study in HRI.

3.6 Obedience

Obedience is usually defined as a form of social influence where a person follows direct orders of a person with authority (e.g. Milgram, 1963). The term obedience is closely connected with Milgram's studies on obedience which are the most important, most well-known and most controversial studies in social psychology (Blass, 1999; Haslam & Reicher, 2017). Their influence is reflected in the immense variety of studies on obedience in many disciplines (e.g. Burger, 2009; Chazan, 2010; Christophe, Bornot, Amado, & Blanc, 2010; Millard 2015; Slater et al., 2006). The following sections present Milgram's studies and the consequences they brought with them in terms of ethical debates. Factors influencing obedience are illustrated with a special focus on empathy in relation to obedience. Furthermore, an overview of research on obedience in HRI is given. The section ends with considerations on the robustness of Milgram's findings over time.

3.6.1 Milgram's Obedience Study

Milgram (1963; 1974) conducted several studies on obedience, though the most well-known study is experiment five against which all other variations are compared. The following description is drawn from Milgram (1963).

40 participants, ranging from 20 to 50 years, with diverse occupations were recruited. There was a confederate, 'Mr. Wallace', who always ended up as the 'learner' and the participant as the 'teacher' in a rigged draw for their role. Fur-

thermore, there was also an 'experimenter', another confederate, dressed in a lab coat. The learner was seated in another room and wired with electrodes. The teacher was instructed to give the learner an electric shock every time the learner made a mistake on a learning task. To persuade the teacher the electric shocks were real, they were demonstrated to the teacher. He then left the room and was seated in another room. Thereafter, unbeknownst to the teacher, the electric shocks were not real.

The teacher was required to give the learner increasingly severe electric shocks – starting from 15, labelled 'slight shock' on the shock machine, rising 30 levels to 450 volts, labelled 'danger – severe shock'. The learner's reactions to the electric shocks were taped and included verbal as well as nonverbal protests increasing in intensity. At 300 volt the learner pounded on the wall and gave no further response. The experimenter, located in the same room as the teacher, gave four 'prods' when necessary:

Prod 1 – 'Please continue' or 'Please go on'
Prod 2 – 'The experiment requires that you continue'
Prod 3 – 'It is absolutely essential that you continue'
Prod 4 – 'You have no other choice, you must go on'.

12.5% of participants stopped at 300 volts ('intense shock') and 65% continued to the highest level of 450 volts. Milgram (1963) also gave anecdotal evidence of behavioral signals of participants: Many seemed to "sweat, tremble, stutter, bite their lips, groan, and dig their fingernails into their flesh" (Milgram, 1963, p. 375). Interestingly, prior to the study, 14 students predicted that less than 3% of 100 persons placed in the experimental setting would continue to 450 volts, which was almost the same as predictions from 40 psychiatrists (Milgram, 1963; 1965b).

3.6.2 Impact of Milgram's Obedience Studies

Milgram's studies raised many concerns, especially about ethical issues (see 3.6.2), but also, among others, on aspects concerning internal validity and theoretical limitations (for a critique regarding theoretical limitations see Haslam & Reicher, 2017). Concerning internal validity, Orne and Holland (1968) as well as Perry (2013) argued that participants did not really believe the electric shocks were real. This is also supported by listening to the original participants' recordings where many expressed their doubts (Perry, 2013). However, Milgram (1963; 1974) reported that 70% of his participants believed the shocks were genuine.

Furthermore, a host of conceptual replications prove the authenticity of the out-
comes of Milgram's observations (e.g. Blass, 2004; Burger, 2009; Doliński et al.,
2017; Slater et al., 2006). The observed behavioral reactions that indicated participants' stress, also suggest that the experience was perceived as very real (Haslam,
Reicher, & Millard, 2015). More recent replications of Milgram's findings include, for example, a French television documentary where participants were led
to believe they were contestants in a pilot episode for a new game show. The pre-
senter ordered participants to give (fake) electric shocks to other participants –
who were in fact confederates – in front of a studio audience. The grand majority
of participants (80%) gave the maximum shock of 460 volts to an unconscious
participant (Le Jeu de la Mort, The Game of Death, 2010: Christophe et al., 2010;
Chazan, 2010). The findings prove the robustness of Milgram's findings, even
more than five decades later (see also Haslam & Reicher, 2017).

Milgram's research was among one of those studies that have raised ethical
concerns and made it a priority for psychology. The psychological harm to the
participants is one of the main concerns but also other aspects are of importance.
Baumrind (1964) for example criticized the deception used in Milgram's study.
Not only were participants betrayed in their trust but the reputation of psychologists and their research also suffered. Nowadays, all professional psychological
associations publish guidelines for ethical conduct that include, for example, to
inform participants of their right to withdraw, to obtain a fully informed consent
and protect participants from the risk of psychological and physical harm (see
e.g., code of ethics of the Deutsche Gesellschaft für Psychologie [German Psychological Association], 2018b). Due to concern for the participants' well-being,
research on obedience to authority has stagnated (Blass, 1999; Blass, 2009; Elms,
2009). Adaptations to the original experiments are necessary when conducting
research on obedience (e.g. Slater et al., 2006). Geiskkovitch, Cormier, Seo, &
Young (2016) also address this issue and the challenges that confront research
ambitions in the context of obedience to authority in HRI (see also Cormier et
al., 2013).

3.6.3 Factors Influencing Obedience

In his diverse variations, Milgram (1965b, 1974) identified several aspects that
influence obedience rates. The focus is laid on those more relevant to this thesis.

3.6.3.1 Situational Variables

In his diverse variations of studies on obedience Milgram identified several factors related to the external circumstances rather than to dispositional variables: proximity, location and uniform (Milgram, 1974).

Proximity with the learner. In Milgram's most well-known study, the teacher and learner are separated from each other; the teacher could only hear the learner but did not see him. The obedience rate of those who fully obeyed was 65%. It dropped to 40% after moving the learner into the same room with the teacher. And further decreased to 30% when the teacher was ordered to force the learner's hand onto an "electroshock plate" (Milgram, 1965b; 1974).

Proximity with the experimenter. In a variation of the original experiment, the experimenter left the room where the teacher gave electric shocks and instead gave the teacher instructions by phone. Obedience rates further dropped to 20.5% (Milgram, 1974).

Location. A comparison of obedience rates between the Yale University as setting for the experiment and a run-down office showed a decrease. 47.5% of participants still fully obeyed when the experiment took place in a run-down building (Milgram, 1974).

Uniform. When the experimenter wore everyday clothes rather than a grey lab coat and represented an "ordinary member of the public" (a confederate) obedience rates dropped to 20% (Milgram, 1974).

3.6.3.2 Legitimation of Authority and Expert Knowledge

Research has shown that legitimation of authority has an influence on obedience (e.g., Blass & Schmitt, 2001; Geiskkovitch et al., 2016; Milgram, 1963). People are more likely to obey other people who they perceive have authority over them. This authority is generally justified by the individual's position of power within a social hierarchy (e.g. Milgram, 1963, 1974).

The perception that someone is in a legitimate position of authority does not need many cues: people expect certain situations to have a socially controlling figure (Milgram, 1974). Thus, the experimenter only needs "a few introductory remarks" together with confidence and an "air of authority" (Milgram, 1974, p. 139) and participants perceive a match between their expectations to find an authority figure and the experimenter who fills this gap and his position is thus not challenged (Milgram, 1974).

Milgram (1974) stated that "the power of an authority stems not from personal characteristics but from his perceived position in a social structure" (p. 139) and that "an authority system (…) consists of a minimum of two persons sharing the expectation that one of them has the right to prescribe behavior for the other" (pp. 142-143). The experimenter is thus seen as one who has a right to issue commands. There are, however, different views challenging the notion that legitimacy is the single reason for obedience. Morelli (1983) rather views the experimenter as an expert authority: Participants obey the experimenter not because he is in charge, but because he has expertise on the topic. This view is shared by several authors (e.g. Greenwood, 1982; Penner, Hawkins, Dertke, Spector, & Stone, 1973). Milgram himself, in a later account, admitted that "both components co-exist in one person. The experimenter is both the person 'in charge' and is presumed by subjects to possess expert knowledge." (Milgram, 1983, pp. 191-192).

Evidence supporting that legitimacy of authority as well as expert authority might have been the cause for obedience comes from Blass and Schmitt (2001). Students identified the experimenter as responsible for the harm done to the learner after watching a film of Milgram's study (the documentary film 'Obedience', Milgram, 1965a). They also indicated that legitimate authority as well as expert authority was the reason for the attribution of responsibility. The authors conclude that both factors combined were the cause for obedience to the experimenter (Blass & Schmitt, 2001; Burger, 2009; Milgram, 1983). Milgram also pointed out that both factors, position and expertise, of an authority figure often occurs in real-life (Milgram, 1983).

3.6.3.3 An "Assistant" Giving Orders

Milgram (1974) varied several aspects in his studies on obedience and one of it was removing the experimenter's legitimate authority by replacing the experimenter with an ordinary man. The ordinary man, in truth a confederate, appears to be a participant in the experiment with the task of recording times. A rigged telephone call that draws the experimenter away from the laboratory presents an opportunity for the other subject to take over a role that could be described as "assistant of the experimenter": Since the experimenter only stated that the 'teacher' should continue with the experiment but without indicating which shock levels are to be used, the confederate suggests to increase the shock level each time the learner makes a mistake. The confederate insists on this procedure throughout the experiment (Milgram, 1974). "Thus, the subject is confronted with a general situation that has been defined by an experimental authority, but with orders on specific levels issued by an insistent, ordinary man who lacks any status as an authority" (Milgram, 1974, pp. 93ff). Obedience rates dropped sharply: 80% of participants were disobedient (Milgram, 1974).

3.6.3.4 Autonomous vs. Remote-Controlled Robot

Research on robots in authority positions is scarce. A study by Cormier et al. (2013) uses a robot as experimenter and assesses participants' obedience rates. The robot Nao was introduced as *"highly advanced in artificial intelligence and speech recognition"* (Cormier et al., 2013, p. 4). The authors also report several nonverbal behaviors implemented in the robot, such as speaking with a neutral tone, gazing around the room and using empathic hand gestures (Cormier et al., 2013). In another study, the Nao robot was introduced as being remote-controlled and thus mirrored the Milgram (1974) experiment where the experimenter communicated by telephone (Geiskkovitch et al., 2016). In Milgram's experiment seven, as the experimenter was physically removed from the laboratory, obedience rates dropped to 20.5%.

3.6.3.5 Dispositional Explanations

Even though only few studies on obedience have used personality measures, some findings suggest an effect of personality on obedience (Blass, 1991). However, there is also evidence that personality might not play a role in obedience behavior (e.g. Burger, 2009).

Especially trait empathy is considered to play a role in obedience (Burger, 2009). Affective sharing between the self and the other is among the main processes involved in empathy (Decety & Jackson, 2004). Affective sharing of another's pain can either lead to sympathy or to personal distress, i.e. either concern for the other or to a self-focused emotional reaction (Decety & Lamm, 2009). Moral reasoning and altruism is associated with empathic concern (Batson et al., 1991) whereas personal distress goes along with a desire to lift the own sorrows (Batson, 1991). Trait empathy should therefore be involved in in the dilemma between obedience to authority and empathy with the victim. However, findings are contradictory: Even though participants who were high in empathic concern expressed reluctance to continue, it "did not translate into a greater likelihood of refusing to continue" (Burger, 2009, p. 10). An association between negative affect and obedience could be found: those who reported high levels of negative feelings were more reluctant to obey (Zeigler-Hill, Southard, Archer, & Donohoe, 2013). Cheetham, Pedroni, Antley, Slater, and Jäncke (2009) also found that personal distress and fantasy (the tendency to transpose oneself imaginatively into the feelings and actions of fictional characters; Davis, 1983) predicted brain activity while watching a virtual "learner" in pain.

Although slightly favoring the explanation that situational variables have more power than dispositional factors (e.g. Milgram, 1963, 1965b, 1974; Burger, 2009), the findings are still contradictory and require further consideration.

3.6.3.6 Gender Differences

Do men and women differ in obedience rates? Milgram (1974) and other authors have analyzed gender differences in (partial) replications and variations of Milgram's obedience studies and found no effect of gender on obedience rates (e.g. Blass, 2000; Burger, 2009; see Sheridan & King, 1972 for an exception). However, studies show that women reported higher levels of nervousness and tension (Burger et al., 2011) as well as higher levels of stress (Milgram, 1974). Findings are inconsistent and seem to occur mainly for self-reported feelings. Hence, this remains to be further investigated.

3.6.3.7 Questionnaires vs. Laboratory

Milgram described the experimental setting of the obedience study and asked students as well as colleagues and psychiatrists what they thought a) they themselves and b) how many participants would administer the maximum shock level. It was predicted that only an insignificant minority would continue to 450 volts and that they themselves would not continue (Milgram, 1963; 1965). However, in the laboratory setting, 65% continued to 450 volts (Milgram, 1963). This shows clearly that people's opinions about their own or others' behavior regarding such a setting was at least far from reliable in being able to predict people's real behavior.

3.6.4 Obedience vs. Empathy

Milgram's obedience studies derive most of its drama from the tension between the demands of the experimenter and the victim, which makes the paradigm psychologically intense (Millard, 2014; see also Milgram, 1963). Milgram himself mentions that "(...) the conflict stems from the opposition of two deeply ingrained behavior dispositions: first, the disposition not to harm other people, and second, the tendency to obey those whom we perceive to be legitimate authorities" (Milgram, 1963, p. 378). It thus seems reasonable to argue that participants are more likely to stop the experiment if empathy with the victim is more powerful than the wish to obey the experimenter. Participants with higher empathic concern did indeed express reluctance to continue the experiment earlier than those with lower empathy level (even though it did not have an effect on obedience rates; Burger, 2009; see also section 3.6.3.5). Even more effective in terms of decreased obedience rates seems to be the presence of the learner as well as the direct contact with the victim in comparison to a distant victim (Milgram, 1965b) (see also section 3.6.3.1).

3.6.4.1 Empathic Reactions

The Milgram-experiment is a social dilemma between obedience to the person with authority and empathy with the victim. Studies show that empathy with the victim is associated with reluctance to continue the experiment (Burger et al., 2011). Even though the majority of participants obeyed an authority figure in obedience experiments, many expressed signs of discomfort, frustration, stress and empathy. Milgram (1965b) points out that the victim's suffering, especially when he is in the same room as the participant, triggers empathic responses. He reasons that "diminishing obedience, then, would be explained by the enrichment of empathic cues" (Milgram, 1965b, p. 63). In order to reduce discomfort, participants looked away in shame or embarrassment (Milgram, 1965b). Slater et al. (2006) also report that participants were "stressed by the situation" (p. 2). Results from physiological measurements confirm the greater overall arousal of participants, especially when they saw the virtual learner's pain (Slater et al., 2006). These reactions were observed in spite of the artificiality of the victim's "pain". Most other studies also report anecdotal evidence of empathic reactions in obedience situations (e.g. Bartneck & Hu, 2008; Burger, 2009; Geiskkovitch et al., 2016).

3.6.4.2 Hesitation Time

There is evidence that the concern for the victim in obedience studies is expressed by a delay in administering electric shocks. Slater et al. (2006) found that those who saw a virtual learner's pain waited much longer before giving an electric shock than those who communicated only through a text interface. They interpreted this finding as care for the well-being of the virtual human (Slater et al., 2006). In a study by Horstmann et al. (2018), participants waited longer to switch off a robot that voiced an objection to being switched off. Sheridan and King

(1972) also found a greater resitancy to hurt the victim (decrease in duration time on the shock button as the shocks got higher). In a reanalysis of a partial replication of Milgram's obedience study Burger, Girgis, and Manning (2011) reported that participants who expressed a concern for the learner's well-being also demonstrated a greater reluctance to press the shock buttons. A certain "reluctance to administer shocks beyond the 300-volt level" (p. 376) is also reported by Milgram (1965b).

3.6.5 Obedience in a Virtual Reality Setting

One of the replications more relevant to the present thesis is a study conducted in a virtual reality setting, replacing the human "learner" in Milgram's original experiment (1963) by a virtual human. Similar to the original experiment, participants were told to give electric shocks to a virtual human "learner" for every incorrect answer (Slater et al., 2006). Participants saw her "pain" and protests. Even though the virtual context implied no real pain for the virtual human straight from the beginning, participants reacted to the virtual person just like towards a real person (Slater et al., 2006). Thus, participants experienced high levels of stress (measured with physiological methods, self-reports of physiological symptoms as well as anecdotal behavioral observations by the experimenters) when confronted with the virtual human's pain compared to only text-based interaction (Slater et al., 2006).

In most prior studies on obedience, deception played a great role. But what is the case when the situation is clearly not real? The learner does and cannot experience real pain? Slater et al. (2006) confronted participants with a virtual human that 'expressed pain' like a real human. And even though the "virtual learner could never be confused with a real human" (p. 6), participants showed strong responses and the authors concluded that "people do tend to respond to the situation as if it were real" (Slater et al., 2006, p. 6), thus confirming the Media Equation approach. In general, research has shown that people tend to treat virtual agents as well as robots as social actors (e.g. Rosenthal-von der Pütten et al., 2013; Menne & Lugrin, 2017, Menne, Schnellbacher, Schwab, 2016; Menne, & Schwab, 2018, Slater et al., 2006; Young et al., 2011).

3.6.6 Obedience in HRI

3.6.6.1 Relevance

Can robots push people to do bad things? Even though obedience is a well-stud-
ied area of psychology, little is known about obedience towards robots. While
obedience to robots might initially sound strange, robots can and are already
placed in authoritative situations (e.g., the military). Obedience to robots is also
relevant where vulnerable people are concerned, e.g., the use of robots in search
and rescue situations, schools and hospitals. Humans often treat robots as social
entities (Bartneck, Verbunt, Mubin, & Al Mahmud, 2007; Kahn et al., 2012;
Menne & Lugrin, 2017; Menne, Schnellbacher, Schwab, 2016; Menne, & Schwab,
2018; Rosenthal-von der Pütten et al., 2013; Rosenthal-von der Pütten et al., 2014;
Short, Hart, Vu, & Scassellati, 2010). And some even proclaim to protect robots
from morally unacceptable behavior (Whitby, 2008; see Thimbleby, 2008 for a
critical juxtaposition). Thus it is important to understand the issue of robotic authority and develop an understanding of determinants and risks of robots in authority positions.

Due to the lack of systematic studies in the context of obedience in HRI, firm
bases for determining the factors and their contributions to obedience are
needed. This is particularly important for HRI since robots are already involved
in military actions (e.g. robot Atlas, Boston Dynamics, 2018). It is also conceiv-
able that robots will take over roles associated with a certain air of authority. A
variety of technical devices are already playing an important role in private life.
If these technical devices are able to show a minimum of social cues, an audio
device like "Alexa" (see e.g., López, Quesada, & Guerrero, 2018, for an overview
on speech-based user interfaces) should be able to give orders. This happens still
in the form of appointment reminders, that is, goals that people are positive
about. However, there are also other developments possible that could turn into
social dilemmas, e.g. one of the simplest the technical device (e.g., Alexa or a social robot) could say would be "The pet is disturbing my reception. You have to
lock it out of the room" or much worse. In order to know how to adequately react
to those developments, or avoid the negative effects of obedience, research into
the conditions of obedience, especially in the field of social robotics, is indispens-
able. However, research in this field is rather scarce and often without systematic
experimental variation. The same is true for the research situation in the entire
field of HRI for that matter (Eyssel, 2017). Psychology can make an important
contribution to the systematic investigation on social and emotional phenomena
in HRI.

3.6.6.2 Obedience to Authority in HRI

While obedience to authority is a well-established research area in psychology, little is known about obedience to robots. Even though research suggests robots are treated as social actors (e.g., Menne & Schwab, 2018) the extent to which this happens still remains unclear. After all, robots are machines, though with a life-like agency (Young et al., 2011). Does authority status play a role in obedience to robots? Is it even possible to perceive a robot to have authority over someone? Would people blame an autonomous robot for their actions performed under its order? Is obedience to a robot more powerful than empathy with a robot? These questions point to the necessity of research in the context of obedience to robots.

Much of the current research in HRI brushes the field of obedience research. Though most studies do not investigate a robot ordering participants to do acts that are morally unacceptable or harmful, the influence of a robot on achieving aims that are desirable is largely investigated.

Research on persuasive robots shows that robots have an influence on psychological factors (e.g. compliance). Chidambaram, Chiang, & Mutlu (2012) showed that a robot that used nonverbal cues could persuade participants into modifying their answers. For example, there is research on how persuasive robots can affect performance in team settings (Liu, Helfenstein, & Wahlstedt, 2008), completion of exercises in rehabilitation therapy (Matarić, Tapus, Winstein, & Eriksson, 2009), or reduction of energy consumption (Ham & Midden, 2014).

More relevant to the present thesis are the following studies where a human or robot experimenter pressures participants into doing uncomfortable tasks. In an experiment by Menne (2017) the robot Nao ordered participants to complete several tasks increasing in psychological effort (from "take a sip of water" to "say something really insulting to me"). Even though sometimes ridicule ("imitate an ape with your hands, feet and sounds"), 77% of participants obeyed the human experimenter and 76% the robot experimenter. Participants also reported feeling slightly more ashamed after the experiment than before, independent of the type of experimenter (human or robot). This exploration into obedience shows that participants generally obey a robot (almost like a human authority) even if the tasks are partly embarrassing or pointless. An experiment based on Milgram's (1963) original obedience study with a robot as victim was conducted by Bartneck and Hu (2008). A human experimenter ordered participants to administer shocks to a robot if it made a mistake in a learning task. The robot uttered verbal and nonverbal protests. Bartneck & Hu (2008) report that even though participants showed compassion for the robot all continued until the maximum voltage was reached. In another study, participants had to "kill" a crawling microbug robot (Bartneck & Hu, 2008). Again, participants expressed sympathy but did as they were told (Bartneck & Hu, 2008). Another study by Bartneck, Bleeker, Bun, Fens, & Riet (2010) investigates to what extent people obey a robot to do embarrassing tasks. In a fake medical examination, participants removed their clothing and put a thermometer in their rectum as the robot ordered them to do so (Bartneck et al., 2010), showing the power a robot might have over humans.

Obedience to robots in an authority position has been explored by Geiskkovitch et al. (2016). There were several conditions: First, the authors compared a human with an autonomous robot (Nao). Second, the autonomous robot (Nao) was compared with a remote-controlled robot (Nao). Third, the autonomous robot was compared with different robot embodiment variants (a computer server vs. a disc-shaped robot). Either the robot Nao or the human experimenter or the different robot embodiment variants directed participants to perform a tedious task: changing extensions on data files manually. Overall, 45% of all participants obeyed the robot experimenter and renamed files for 80 minutes. The authors report that this obedience rate is similar to those found by Milgram and subsequent studies (Geiskkovitch et al., 2016). 86% obeyed the human experimenter, 46% the autonomous humanoid robot. The authors did not find a difference between the autonomous and remote-controlled conditions (Geiskkovitch et al., 2016). The perceived legitimacy of authority of Nao was also assessed, though not as part of the experimental manipulation but post-hoc. Participants who rated the robot as legitimate authority "obeyed less, protested earlier and protested more often" (Geiskkovitch et al., 2016, p. 14). While renaming data files participants showed signs of frustration such as grunting or laughing (Geiskkovitch et al., 2016). The authors conclude that robots can pressure participants to continue a task they do not want to do (Geiskkovitch et al., 2016).

3.6.7 Robustness of Obedience Over Time

The wide-known popularity of Milgram's obedience experiments has raised the question whether the same obedience pattern could still be observed today. Gergen (1973) reasoned that later studies should find less obedience than earlier ones due to knowledge about the unexpected power of authority. Even though ethical concerns prevent full replications of obedience studies, partial replications and variations have found no evidence for a change over time (e.g. Blass, 2004; Burger, 2009), indicating that situational factors still present a powerful force.

3.7 Summary of the Theoretical Background

Section 1 of the theoretical background emphasized that social robots should be able to express readable signs, especially by using natural cues, and exhibit competent behavior. Focusing on emotional expressiveness in robots, examples in research illustrated the power of affective nonverbal cues. Section 2 then introduced the concept of emotions and presented emotion research in psychology and HRI. The passage showed the wide array of different definitions, concepts, models and measurements of emotions and highlighted challenges of emotion research in the field of social robotics. Empathy, closely related to emotions, was introduced in section 3. Different definitions as well as differentiation from similar terms were given. It was shown that even though gender and individual differences in empathic behavior exist, findings are still inconsistent. Methods for measuring empathy are largely the same as for emotional reactions in general. Research on empathic reactions in social robotics research faces many challenges, especially concerning difficulties and inconsistencies in the measurement of empathic reactions. Section 4 presented facial expressions as the most frequently studied nonverbal communication channel. Research in psychology has found evidence for a reliable association between spontaneous facial expressions and subjective experiences. Furthermore, the section showed that research on facial expressions of emotion towards robots remains scarce. Section 5 then emphasized the importance of measurement methods of emotions. Self-reports, observational and physiological methods were introduced including their advantages and disadvantages. The section concluded that there is no gold-standard method for measuring emotions and advocated the use of a multi-method approach. Furthermore, online research versus 'traditional' laboratory research as general research practices were compared. As Experiment 2 and Experiment 3 were dedicated to explore obedience and empathy towards robots, the final section of the theoretical background introduced Milgram's studies on obedience, their impact and different factors, such as situational variables, authority status, dispositional factors, gender differences as well as research method (online vs. laboratory) influencing obedience. Empathic reactions observed in obedience studies were presented as well as obedience settings in virtual reality and HRI. The section showed that humans obeyed robots similar to humans and showed signs of empathic behavior. However, it was illustrated that research in HRI remains scarce and the present thesis is one of the first to systematically explore obedience and empathy towards robots.

3.8 Research Questions

There is one major research question (RQ) encompassing all other objectives of this doctoral thesis, which is:

RQ: Are social robots able to evoke emotional reactions on a subjective level (self-reports) as well as on a motoric-expressive level (facial expressions)?

In other words: How profound are emotional reactions towards robots? Are emotional reactions also observable in the user's face? Embedded in this question, this doctoral thesis aims to answer three main research questions:

RQ₁: Does the appearance, emotional expressivity and treatment of a robot influence emotional reactions in a video-based setting?

RQ₂: Does the text-based description of an obedience scenario in HRI influence emotional reactions?

RQ₃: Does the live interaction with robots in an obedience scenario influence emotional reactions?

In the following sections, these RQs will be explored in three experimental studies.

4 Experiment 1: Emotional Reactions Towards Social Robots

4.1 Study Outline and Hypotheses

The first experimental study has several aims: first, to investigate the profoundness of emotional reactions towards robots. Second, to analyze if there are visible changes in the face in response to a robot being treated in different ways. Third, to study whether there is a potential match between facial expressions and self-reports of emotional states. Fourth, to understand whether a robot's capability to express emotions has an influence on participants' emotional reactions and fifth, if the latter depends on the appearance of the robot. By experimentally manipulating several aspects (emotional expressivity, appearance of the robot, treatment) and using a multi-method approach, this study aims to contribute further to systematic research in the field of social robotics as well as to get a more profound understanding of emotional reactions towards social robots.

Research (see 3.1.1, 3.1.2, 3.2.5.2, and 3.3.5) has shown that people respond emotionally towards an emotionally expressive robot (e.g., Leite et al., 2008; Leite et al., 2010; Menne & Schwab, 2018; Menne & Lugrin, 2017; Pereira et al., 2011; Rosenthal-von der Pütten et al., 2013; Rosenthal-von der Pütten et al., 2014; Salem et al., 2011). And in direct comparison with a non-affective robot, there is evidence that users preferred the interaction with an expressive robot (Bartneck, 2003) or responded earlier and moved faster (Moshkina, 2012). Furthermore, it could be shown that participants feel more positive after watching a robot in a friendly interaction and more negative after watching a robot being mistreated (e.g., Menne & Schwab, 2018; Menne & Lugrin, 2017; Rosenthal-von der Pütten et al., 2013; 2014). As described in section 3.3.5, Riek et al. (2009) reported that participants felt more sorry for a human-like robot than mechanical-looking robots. Furthermore, Rosenthal-von der Pütten and colleagues (2013; 2014) as well as Menne & Schwab (2018) showed that participants empathized with a robot dinosaur. Additionally, emotional responses are reported for an anthropomorphic robot Reeti (Menne & Lugrin, 2017).

As outlined in earlier sections (3.2.5.2, 3.3.5, and 3.4.4.2), systematic research on emotional reactions towards robots remains scarce and hence, it is difficult to formulate a specific hypothesis on the difference between the anthropomorphic robot Reeti and the animal-like Pleo. However, drawing on research outside emotional reactions towards robots, a study by Unz et al. (2008) (see 3.2.5.2) reported participants had more negative feelings after having seen TV news on violence against innocent animals than violence against humans. Taking into account the biophilia hypothesis which assumes "the existence of a fundamental, genetically based, human need and propensity to affiliate with life and lifelike processes" (Kahn, 1997, p. 1), violence against innocent animals should trigger stronger emotional reactions than violence against inanimate objects or even humans. Following this line of thought, as well as findings by Riek et al. (2009), it is assumed that participants will generally react most strongly towards the animal-like Pleo, followed by the anthropomorphic Reeti, compared to the machine-like Roomba, especially if the first ones (Pleo, Reeti) are emotionally expressive and the latter (Roomba) non-expressive (like an inanimate object).

To extend previous research on emotional reactions towards social robots, the effect of appearance (anthropomorphic: Reeti, animal-like: Pleo, and machine-like: Roomba), and emotional expressivity ("on" versus "off") in different scenarios (positive treatment vs. negative treatment) is systematically analysed in Experiment 1.

The following hypotheses are categorized according to the method used to assess the dependent variables. First, hypotheses using self-report measurements are described. Then, those that are based on observational methods are presented and finally, exploratory considerations are introduced.

4.1.1 Hypotheses Based on Self-Report Measurements

Regarding the self-reported emotional state, it was hypothesized that participants will report [Hypothesis] H_{1a}) increased negative feelings after watching the torture video than before, if the robots were emotionally expressive and mostly so for Pleo, followed by Reeti compared to Roomba and H_{1b}) increased positive feelings after watching the friendly video than before, if the robots were emotionally expressive and mostly so for Pleo, followed by Reeti compared to Roomba. Furthermore, participants will report H_{1c}) more negative feelings after the torture video compared to the friendly video, if the robots were emotionally expressive and mostly so for Pleo, followed by Reeti compared to Roomba and H_{1d}) more positive feelings after the friendly video compared to the torture video, if the robots were emotionally expressive and mostly so for Pleo, followed by Reeti compared to Roomba.

Experiment 1 exposed participants to different emotional treatments of robots (friendly vs. torture) and did not leave participants a choice to stop the treatment (low power/control). Following Scherer & Ellgring (2007), participants should therefore H_{2a}) feel more sadness after watching the torture video compared to the friendly video. They will also report H_{2b}) more happiness after the friendly video than the torture video. This will be more so for H_{2c}) emotionally expressive robots than non-expressive robots that are H_{2d}) more animal-like (Pleo), followed by the anthropomorphic Reeti and least for the machine-like Roomba.

For the self-reported evaluation of the videos and the robots (Rosenthal-von der Pütten et al., 2013), it was expected that H_{3a}) participants evaluate the torture videos more negatively than the friendly videos. Especially if the robot was H_{3b}) emotionally expressive (compared to non-expressive) and H_{3c}) less machine-like (compared to the more machine-like robot Roomba). Additionally, participants will report H_{4a}) more antipathy for the non-expressive Roomba than the non-expressive Reeti and least for the non-expressive Pleo. Participants will also report H_{4b}) more antipathy for the emotionally expressive Roomba than for the emotionally expressive robot Reeti and least for the emotionally expressive robot Pleo.

Concerning the self-reported empathy with the robot, participants will report H_{5a}) more pity/ anger at torturer for Pleo, followed by Reeti and Roomba, and H_{5b}) more empathy for Pleo, followed by Reeti, and Roomba after watching the torture videos. This will be more so for H_{5c}) emotionally expressive robots than non-expressive robots. Moreover, it is assumed that participants will attribute H_{6a}) more positive feelings to the emotionally expressive robots than the non-expressive robots in the friendly video, and H_{6b}) more negative feelings to the emotionally expressive robots in the torture video. H_{6c}) the most feelings will be attributed to the animal-like Pleo, followed by the anthropomorphic Reeti, and least to the machine-like Roomba.

4.1.2 Hypotheses Based on Observational Methods (FACS)

This doctoral dissertation chooses to operate on a micro analytic level (AUs) independent of prior assumptions about prototypical emotion expressions. This provides several advantages: First, spontaneous facial expressions often occur in a low intensity and produce only subtle changes in the face (that can change rapidly: micro momentary expressions, e.g. Ekman & Friesen, 2003) which makes it difficult to categorize them as one of the prototypical emotions. Since FACS is not based on categorizing expressions into prototypical emotions, facial behavior is coded more objectively and not dependent on an emotion theory (which is especially favorable considering the lack of a comprehensive emotion theory, see

3.2, and 3.2.4). FACS's sensitivity to detect subtle expression differences were for example demonstrated by Del Giudice and Colle (2007) (differences between fake and genuine smiles; see also Ekman & Friesen, 1982) or Prkachin (1992) (characteristics of painful expression). Second, research has shown that facial expressions of emotion are more often partial than complete (Carroll & Russell, 1997), thus questions of how single AUs and combinations can be interpreted become crucial (Scherer, Schorr, & Johnstone, 2001). Third, using AU codes instead of emotion codes can be useful for testing the validity of emotion specific expression typologies (Olderbak, Hildebrandt, Pinkpank, Sommer, & Wilhelm, 2013). Fourth, in the absence of a comprehensive emotion theory, facial behavior is discussed to be multifunctional (Ekman, 1993; Russell & Fernández-Dols, 1997; Scherer et al., 2001).

As outlined earlier (see 3.4.4), research has shown that the most frequently used facial muscles/ AUs are the lip corner puller (AU 12) and the brow lowerer (AU 4) which are associated with positive emotions and negative emotions, respectively (e.g., Dimberg & Thunberg, 1998). Zygomaticus major (involved in AU 12) and corrugator supercilii (involved in AU 4) have also been activated in response to happy and sad robotic faces, respectively (Riether, 2013; see section 3.4.4.2). Moreover, AUs 1, 2, 4, 5, 7, 9, 10, 11, 15, 16, 17, 20, 22, 23, 24, 25, 26, 27 are considered to occur in fear, anger, sadness, surprise and disgust (Ekman et al., 2002; see also Scherer & Ellgring, 2007), which are regarded as emotions with a negative valence (e.g., Smith & Ellsworth, 1985). In contrast to that, AU 6 and AU 12, occurring in the so-called "Duchenne smile", are associated with feelings of pleasantness and joy (Ekman, Davidson, & Friesen, 1990). Hence, it is assumed that, depending on the emotional expressivity and appearance of the robot, participants will show H_{7a}) more AU12 as well as H_{7b}) more AUs associated with positive emotions (AU 6, 12) in the friendly video. Furthermore, they will show H_{7c}) less AU 12 and H_{7d}) less AUs associated with positive emotions in the torture video. Additionally, participants will show H_{8a}) more AU 4 as well as H_{8b}) more AUs associated with negative emotions (AU1, 2, 4, 5, 7, 9, 10, 11, 15, 16, 17, 20, 22, 23, 24, 25, 26, 27) in the torture video. Likewise, H_{8c}) less AU4 and H_{8d}) less AUs associated with negative emotions in the friendly video. The emotional reactions are assumed to be most strongly for H_{8e}) emotionally expressive robots than non-expressive robots that are H_{8f}) more animal-like (Pleo), followed by the anthropomorphic Reeti and least for the machine-like Roomba.

4.1.3 Exploratory Considerations

As described above (3.3.3), women are typically reported as being more empathic (e.g., Cheng et al., 2006; Eisenberg & Fabes, 1990). However, findings on gender differences regarding emotions and emotional reactions towards robots are inconsistent (see section 3.3.3). Furthermore, the effect of individual dispositions such as affiliative tendency, loneliness and empathy trait (as outlined in 3.3.3) will be explored here.

Regarding the influence of personality traits and the effect of gender, it was assumed that women and participants high in affiliative tendency, loneliness and empathy (trait) will show more emotional reactions in self-report and facial expressions.

4.2 Methods

4.2.1 Participants

Data were collected from 243 participants (31.7% male) with a mean age of 23.4 years ($SD = 8.12$, range = 16 - 62) using the internal recruitment system of the Institute Human-Computer-Media at the University of Würzburg. Participation was on a voluntary basis and participants were offered course credit. The grand majority of participants (70.4%) were unfamiliar with social robots, while 29.6% described they only had superficial experience with robots in general (e.g., working with industrial robots). None of the participants had previously encountered the robots Pleo, Reeti and Roomba. The majority of participants (88%) were highly educated (i.e., university entrance certificate or university degree). The remaining participants had left school with less than thirteen years of formal education. Most participants (84%) were undergraduate students enrolled in different degree programs (e.g., media communication, biology, sports), followed by employees (9.9%), self-employed (2.1%), school students (1.2%) and other (2.8%). 81 participants were in the Pleo condition, 84 in the Reeti condition and 78 in the Roomba condition. Due to technical problems such as failure in video recording or network failure, 10.3% of participants had to be discarded for the analysis of the video data. That left 75 participant data points appropriate for FACS coding in the Pleo condition, 74 in the Reeti condition and 69 in the Roomba condition. Written informed consent was obtained from each participant prior to the study, in line with a protocol approved by the Ethical Committee of the *Deutsche Gesellschaft für Psychologie* (2018a, 2018b) (German Psychological Association).

4.2.2 Stimulus Material

4.2.2.1 Robots

Pleo. The entertainment robot Pleo (Innvo Labs, 2012) is 20 cm in height and 50 cm in length. Its appearance is inspired by a baby Camarasaurus. In an interview with the sales director of Pleo, the reason for this design choice was explained: no expectations regarding the realism of movement or appearance are elicited by using the shape of a dinosaur instead of, for example, the shape of a cat or a dog (Pluta, 2008). The robot is built to be "a life form" (Innvo Labs, 2012), that "thinks and acts independently, just like a real animal" (Innvo Labs, 2012) and is thus intended to be used as an entertaining companion in everyday life. Associations with real animals (pets) are intended since Pleo does react autonomously to its environment. Sensors placed all over its body allow it to react to touch, movement, temperature and light. Through a speaker, Pleo is able to make utterances. Pleo is able to express emotions like joy and pain through noises and movement. It was chosen for the present study due to its non-threatening, animal-like, small size, and its affordability. Furthermore, it does not require programming skills.

    Roomba. The "IRobot Roomba 774 Vacuum Cleaning Robot" (iRobot, 2019) is a service robot, "powered by a full suite of smart sensors that automatically guide the robot around" (iRobot, 2019). It is 9.2 cm in height and 35 cm in diameter. Its interface contains several control buttons to set time, day and duration of cleaning. When Roomba is switched on, its "clean" button turns green and the robot begins cleaning. There are no other sounds during cleaning except the usual noises of a vacuum cleaner with 60 dB. Roomba moves forwards by turning the wheels attached to its bottom and vacuums the floor with brushes. To change direction, the robot can rotate on its own axis. Roomba is able to detect obstacles like edges or stairs that prevents the robot from falling down. Roomba's disc-shaped appearance and its neutral design evokes associations with a machine rather than with a living being. Due to its small size, its round features and neutral design Roomba looks harmless and was chosen for Experiment 1. Furthermore, Roomba does not require further programming skills, it "reacts" automatically.

    Reeti. In the strictest sense, Reeti is a PCBot, a mixture between a Personal Computer and a robot, since it is constituted of a computer in its body and a robotic head (Robopec, 2015). As Reeti appears more like a robot than a PC and is mostly called a "robot" by the manufacturers (Robopec, 2019) it will be called "robot" in the following sections. Reeti is 44cm tall (Robopec, 2019). With its movable expressive head and its firmly built body (lacking arms or legs) it is a more anthropomorphic robot than Pleo or Roomba. Reeti's head is equipped with an elastic skin and 15 degrees of freedom to move the ears, the neck, the eyes, eyelids and the mouth (Robopec, 2015). "He expresses in an interactive way his feelings, speaks, sees, and feels the touch" (Robopec, 2019). The cheeks are equipped with coloured LEDs to express emotions and sensors for interaction when touched.

Using speech synthesis and microphones, Reeti converts text input into speech (text-to-speech) or plays recorded audio tracks. DiSalvo, Gemperle, Forlizzi, & Kiesler (2002) made several suggestions for a humanoid robotic head, such as wide head, wide eyes and skin. Due to its cartoon-like appearance, its neutral design (white body), its small size and its constraints in movement (absence of limbs), Reeti appears to be more in need of protection than threatening. Furthermore, due to its appearance, it does not evoke inappropriate expectations

Table 5. Description of the robots used in Experiment 1

Robot	Appearance	Function[7]	Emotional Expressivity „on"	Emotional Expressivity „off"
Roomba	Disc-like, wheeled robotic vacuum cleaner	Vacuum cleaning robot (iRobot, 2019)	Moves; green LED light; vacuum cleaner noises	No movement, lights or sounds
Reeti	Anthropo-morphic robot with an expres-sive face	Expressive and communicating robot (Robopec, 2019)	able to show fa-cial expressions and make utter-ances; LED lights in cheeks	No movement, lights or sounds
Pleo	A small robot di-nosaur	Autonomous robotic life form (Innvo Labs, 2012)	animal-like be-havior (can ex-press emotions like joy and pain through sound and movement)	No movement or sounds

---

[7] As advertised by manufacturer

towards the realism of the robot's behavior and is hence ideally suited for the research goals of Experiment 1.

Reeti's facial expressions, speech recognition and –synthesis as well as the image processing of the robot's eye cameras can be programmed using an Open Source Software (for further information see Robopec, 2015). Since Reeti did not react autonomously to the treatment shown in the videos (see 4.2.2.2), the robot had to be programmed to "react" appropriately to the treatment he received (e.g. moaning and showing facial expressions of unpleasantness while being hit).

Table 5 presents a description of the robots and the different conditions of the robot's emotional expressivity.

4.2.2.2 Video Clips

Videos are commonly used for emotion elicitation (see section 3.2.3). A major focus laid on reliable and replicable robotic behaviors as well as a facilitated recording of participants' facial expressions, which is why videos instead of live interactions with the robots were chosen.

The robots Pleo, Reeti and Roomba were used to create video clips of one minute in length each. The scenes were filmed using the Camcorder Panasonic HDC-SD909 and post-edited using Adobe Premiere. The content and format of the video clips was inspired by the study of Rosenthal-von der Pütten et al. (2013). The authors created video sets of Pleo that contained either a "friendly interaction" or a "torture interaction" and consisted of five sequences each lasting ten seconds, separated by a two seconds black screen. Every video set lasted one minute. These formal settings were also used for the videos created for Experiment 1. However, since special focus was laid on the comparability of the video contents between the different robots, scenes had to be created that worked for all three robots. For example, in the original video by Rosenthal-von der Pütten et al. (2013), Pleo's head was hit on the table. Due to the sensitive sensors in Reeti's head and its considerably higher acquisition costs, it was impossible to execute this treatment. Especially considering that the act still has to look believingly "cruel". Reeti's head could therefore not simply be "connected" with the table with utmost care but treated with more brutality, just like the other robots' heads, to still count as violent behavior and evoke emotional reactions. Hence, several scenes were created that, a) were practicable for all three robots, b) looked believingly enough to be counted as violent actions and c) were comparable between the three robots (see Table 6). Great care was taken only to display a black sleeved arm or black shoe and not the whole human who gave the treatments to the robots except where it could not be prevented. To avoid participants being

Table 6. Scenes of the video clips for positive treatment and negative treatment for all three robots

	Scene 1	Scene 2	Scene 3	Scene 4	Scene 5
Positive Video	Caressing with a massage ball	tickling	feeding	kissing	stroking
Negative Video	kicking	Strangling with plastic bag	Strangling with cable	boxing	punching

Note. The sequence of the scenes was randomized and three different versions of each video set and robot were created to avoid sequence effects

distracted from the content by design quality issues in the video clips, it was made sure that the scenes were well illuminated and were high in audio-visual quality. Regarding the content, a script was created to ensure treatment of the robots in the different conditions did not differ. To counterbalance sequence effects, the sequence of the scenes was randomized and three different versions of each video set (Reeti, Roomba, Pleo x positive vs. negative treatment x emotionally expressive vs. non-expressive) were created, resulting in 36 video clips in total.

Another factor was also the robot's emotional expressivity. As described in Table 5, the three different robots each have their own way of expressing emotions, limited by their design. At the most basic level, all robots moved and "reacted" to the treatment that was given to them (e.g., Roomba drove towards the "food" and "ate" it by cleaning the table; Pleo purred while being caressed, Reeti displayed facial expressions of pain and moaned while being "suffocated" with a plastic bag, etc.). This was called the "emotional expressivity" condition (shortly also called "on" condition). In the "no emotional expressivity"-condition – shortly also called "off" condition, the robots did not move or make any sounds – they did not react to the treatment.

4.2.3 Self-Report Measures

To capture the main variance of emotional experiences (Mauss & Robinson, 2009; Russell & Barrett, 1999; Watson, 2000), the PANAS questionnaire, a well-established self-report measurement to capture dimensional aspects of emotional states (Weidmann et al., 2017) was used. Second, to supplement the wide variety of emotional experiences and match it with lay people's ideas of the existence of specific emotional states, the M-DAS (Renaud & Unz, 2006) was employed, a questionnaire with established psychometric criteria and frequently used in media research to capture distinct emotional states (see also section 3.5). Furthermore, to ensure comparability with the study by Rosenthal-von der Pütten et al. (2013), the same questionnaires were used (*Emotional State, Evaluation of the Videos, Evaluation of the Robot, Empathy with the Robot, Attribution of Feelings to the Robot, Affiliative Tendency, Loneliness, Empathy Trait*). However, due to time constraints (it was made sure that participants were not unduly burdened for an excessively long time) as well as due to theoretical considerations (see 4.1), subscales of questionnaires were used where indicated. Cronbach's alphas have been calculated based on data of the present study.

4.2.3.1 Emotional State

The german adaptation of the Positive and Negative Affect Schedules (PANAS; Watson, Tellegen, & Clark, 1988) was used. The factor-based scales for "Positive Affectivity" and "Negative Affectivity" contained ten items each. The items were rated on a five-point Likert scale ranging from "nothing or very little" to "very strong". Cronbach's alphas in this study ranged between .802 and .898. Sum scores were calculated for the positive and negative affect subscale for scores prior to the experiment as well as for each video set.

4.2.3.2 Evaluation of the Videos

Rosenthal-von der Pütten et al. (2013) constructed this questionnaire whose items were rated on a five-point Likert scale. Four subscales have been identified by these authors, but only one subscale was used: *Negative Video* (six items; e.g. the movie was disturbing, repugnant, etc.; Cronbach's $\alpha_{positive\ video}$ = .771, Cronbach's $\alpha_{negative\ video}$ = .895). Sum scores were calculated.

4.2.3.3 Evaluation of the Robot

This questionnaire, using a seven-point Likert scale, was also designed by Rosenthal-von der Pütten et al. (2013). The authors found five factors, however, only one scale was used: *Antipathy* (three items; e.g. unlikable, cold, etc; Cronbach's $\alpha_{positive\ video}$ = .816, Cronbach's $\alpha_{negative\ video}$ = .727). Sum scores were calculated.

4.2.3.4 Empathy with the Robot

Rosenthal-von der Pütten et al. (2013) designed this questionnaire, containing two subscales rated on a five-point Likert scale: *Pity for robot/Angry at torturer* (five items; e.g. I felt pity for the robot; I hoped that this treatment would stop soon, etc; Cronbach's $\alpha_{\text{positive video}}$= .524, Cronbach's $\alpha_{\text{negative video}}$= .766). *Empathy with the robot* (seven items; e.g. I sympathized with the robot's situation; The robot did not feel anything [reverse coded], etc; Cronbach's $\alpha_{\text{positive video}}$= .866, Cronbach's $\alpha_{\text{negative video}}$= .886). Due to low internal consistency, the subscale Pity for robot/Angry at torturer was not used for the positive video. Sum scores were calculated.

4.2.3.5 Attribution of Feelings to the Robot

This scale was taken from Rosenthal-von der Pütten et al. (2013) and contained ten items rated on a five-point Likert scale (e.g., I can imagine that: ...the robot had fun, the robot was very relaxed, etc.; Cronbach's $\alpha_{\text{positive video}}$= .849, Cronbach's $\alpha_{\text{negative video}}$= .833). Sum scores were calculated. The authors also mention that "high sum scores indicate the attribution of positive feelings to the robot whereas low sum scores indicate the attribution of negative feelings to the robot" (p. 25).

4.2.3.6 M-DAS

A modified version based on the Differential Emotion Scale (DES; Izard, Dougherty, Bloxom, & Kotsch, 1974), the Modified Differential Affect Scale (M-DAS; Renaud & Unz, 2006), was used to assess a broader range of subjective emotional states. It contains distinguishable emotion categories such as pleasure, happiness, sadness, anger etc. (48 items) rated on a five-point Likert scale. Participants completed the whole questionnaire, but for the present study, only the subscales *Happiness* (Cronbach's $\alpha_{\text{positive video}}$= .851, Cronbach's $\alpha_{\text{negative video}}$= .786) and *Sadness* (Cronbach's $\alpha_{\text{positive video}}$= .566, Cronbach's $\alpha_{\text{negative video}}$= .791), were used due to theoretical considerations (see 4.1). Sum scores were calculated. Subscales, such as *Sadness*, were presented for both videos to ensure uniformity, however, misunderstandings concerning the understanding of certain items cannot be excluded (e.g., the participant was asked if he/she felt discouraged while watching the friendly interaction with the robot). These misunderstandings might now be reflected in the low internal consistency scores. Due to the low internal consistency of the subscale *Sadness* for the positive video it was excluded from further analyses.

4.2.3.7 Affiliative Tendency

Participant's affiliative tendency was assessed using the german version (Teubel, 2009) of the Mehrabian Affiliative Tendency Scale (Mehrabian, 1976).The scale consists of 26 items (e.g. "I would rather express open appreciation to others most of the time than reserve such feelings for special occasions"; "Having friends is very important to me") rated on a nine-point Likert scale (Cronbach's α = .796). The affiliative tendency scale "measures an individual's general expectation of the positive reinforcing quality of others" (Mehrabian, 1970, p. 417). People scoring high on affiliative tendency are generally sociable, friendly, and open about their feelings. Sum scores were calculated for participants' affiliative tendency.

4.2.3.8 Loneliness

The most widely used instrument for assessing loneliness is the UCLA Loneliness Scale. The german translation (Lamm & Stephan, 1986) of the revised UCLA Loneliness Scale developed by Russell, Peplau, and Cutrona (1980) was used. The scale consists of 20 items rated on a four-point Likert scale. Example items are "I lack companionship"; "I feel isolated from others") (Cronbach's α = .908). Sum scores were calculated.

4.2.3.9 Empathy Trait

The Interpersonal Reactivity Index (IRI; Davis, 1983) is one of the most frequently used questionnaires for assessing participants' dispositional empathy (e.g., Paulus, 2009). The german version of the IRI, the "Saarbrücker Personality Questionnaire" (SPF) by Paulus (2009) was used. The IRI and SPF, respectively, consist of a set of four factors, each factor representing another dimension of empathy. The *perspective-taking scale* assesses the self-reported tendency to spontaneously adopt the psychological point of view of others (e.g., "I sometimes try to understand my friends better by imagining how things look from their perspective") (Cronbach's α = .755). The *fantasy scale* measures participants' tendency to transpose themselves into the feelings and actions of fictional characters and fictional situations, e.g., "When I am reading an interesting story or novel, I imagine how I would feel if events in the story were happening to me" (Cronbach's α = .711). The *empathic concern scale* assesses "other-oriented" feelings of sympathy and concern, for example "I often have tender, concerned feelings for people less fortunate than me" (Cronbach's α = .713). The *personal distress scale* measures the self-reported tendency to experience "self-oriented" feelings of anxiety and discomfort in tense interpersonal situations, e.g., "Being in a tense emotional situation scares me" (Cronbach's α = .711). Sum scores were calculated for each subscale.

4.2.4 Behavioral Measures: Facial Expressions

Participants' facial activity was recorded by using the internal webcam of the computer (Figure 1), resulting in 436 minutes of video material in total. One minute of video took between 20 and 60 minutes of time to be coded by a certified FACS coder (a coder who is trained in FACS and has achieved at least 80% inter-rater reliability in the FACS Final Test; cf. Ekman et al., 2002). The videos were first watched in real-time, then in slow motion to detect all Action Units that

Table 7. Action Units and Action Descriptors observed in Experiment 1[8]

Appearance Changes		Appearance Changes	
AU 1	Inner Brow Raiser	AU 20	Lip Stretcher
AU 2	Outer Brow Raiser	AU 21	Neck Tightener
AU 4	Brow Lowerer	AU 22	Lip Funneler
AU 5	Upper Lid Raiser	AU 23	Lip Tightener
AU 6	Cheek Raiser & Lid Compressor	AU 24	Lip Presser
AU 7	Lid Tightener	AU 25	Lips Part
AU 9	Nose Wrinkler	AU 26	Jaw Drop
AU 10	Upper Lip Raiser	AU 27	Mouth Stretch
AU 12	Lip Corner Puller	AU 28	Lips Suck
AU 14	Dimpler	AD 32	Bite
AU 15	Lip Corner Depressor	AD 34	Puff
AU 17	Chin Raiser	AU 39	Nostril Compressor
AU 18	Lip Pucker		

Note. AU = Action Unit. AD = Action Descriptor.

---

[8] Sometimes unilateral AUs occurred: the action occurred on only one side of the face. Hence, the abbreviation "L" for left and "R" for right is placed in front of the AU number (Ekman et al., 2002).

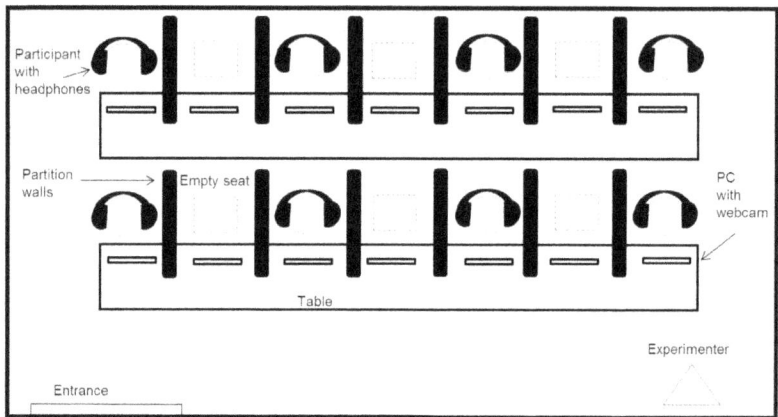

Figure 1. Sketch of the experimental setting of Experiment 1 (source: own figure)

appeared (Ekman et al., 2002). According to Ekman et al. (2002), the perceptual apex of the following AUs was coded (Table 7). The reliability of the coding was proved by independent coders trained in FACS. Interrater reliability for coding was good to excellent (Cohen's κ ≥ .76) according to Sayette, Cohn, Wertz, Perrott, and Parrott (2001). Examples of facial appearance changes observed in Experiment 1 are not printed here to protect participants' personal data.

4.2.5 Procedure

A 3x2x2 factorial mixed design was used with treatment as within-subjects factor and emotional expressivity and type of robot as between-subjects factors. Participants were randomly assigned to one of the six conditions. The experimental procedure took place from June to July 2016 in a computer lab of the Julius-Maximilians-University of Würzburg. A sketch of the experimental setting can be seen in Figure 1. A maximum of eight people could participate in one session. Every session lasted about 30-40 minutes. To minimize the social context of the experimental setting and its possible influence on emotional reactions (e.g., Fridlund, 1992) partition walls were used to isolate participants from each other. Furthermore, one seat was left empty between participants. At least one experimenter was present at all times, however he / she was blocked from view by the partition walls and PC screen to ensure miminum distraction for the participants. In the beginning, written informed consent was obtained from all participants (video- and audio recording, participation in the study according to the German

Psychological Association, 2018a; 2018b). Participants were then instructed to complete a web-based questionnaire that was already opened on a computer screen in front of them and follow the instructions of the online questionnaire. After generating their own code (which was needed to anonymously match the questionnaire data with the video data), participants completed the *PANAS* questionnaire, followed by demographic data. Then, the questionnaires *SPF, loneliness* and *Affiliative Tendency* were completed. After that, participants were instructed to put on the headphones to watch the first video. The video was chosen out of the 36 video clips (see 4.2.2.2). All participants either saw the positive video or the negative video first (the sequence was randomized between participants). Each participant only saw one robot that was either emotionally expressive ("on") or not ("off"). While participants watched the video clips, their facial expressions were recorded using the camera embedded in the PC. After the video, participants completed the *PANAS* scale, a written statement on how they perceived the video, the *M-DAS*, scales *evaluating the video* and *the robot*, as well as the *empathy for the robot* and *attribution of feelings to the robot* scales. Then, the next video was started (the negative video if the positive video was the first and vice versa). The procedure and scales that followed were the same as after the first video. After that, participants were fully debriefed and thanked for their participation.

4.3 Results

4.3.1 Statistical Analyses: Test Assumptions

In this thesis, mixed-design ANOVAs were employed. This type of parametric test requires several assumptions to be met, thus prior to performing statistical analyses, test assumptions were checked. Data were tested for outliers using boxplots and histograms as well as Cook's distance. The assumption of normality is commonly tested using P-P-Plots as well as Shapiro Wilk test. However, t-test and ANOVA are robust against violations of normality in sample sizes ≥ 30 (Bortz, 2005; Kubinger, Rasch, & Moder, 2009). As sample sizes in this thesis were ≥ 30, testing the assumption of normality was not necessary (Field, 2013).

Additivity, linearity, independence of observations and homogeneity of variances was tested using diagnostic plots of estimated residuals. For homogeneity of variances, Levene's test was used additionally. Violations of homogeneity were observed for a part of the analyses in this thesis. However, since cell sizes were roughly equal[9], ANOVA can be considered robust against this violation (Eid, Gollwitzer, & Schmitt, 2010; Hussy & Jain, 2002).

Another assumption in mixed-design ANOVA is sphericity. However, sphericity is only an issue if at least three conditions of the within-group variable were used (Field, 2013). In this doctoral thesis the within-group variable only had two levels and thus sphericity was not a concern. Unless otherwise stated, Šidák correction was used for the simple effects analysis as it is quite similar to the Bonferroni correction but less conservative (Field, 2013). For regression analysis, normality, homogeneity of variances as well as multicollinearity (r < 9) was checked (Field, 2013). Unless otherwise reported, all assumptions were met. For all statistical tests, an alpha level of .05 was set. Data preparation was conducted in Microsoft Excel 2013. For statistical analyses, IBM SPSS (version 23.0) was used.

4.3.2 Design and Statistical Analyses

Study 1 followed a 3 (type of robots Pleo, Reeti, Roomba) x 2 (treatment: friendly interaction vs. torture interaction) x 2 (emotional expressivity: on vs. off) mixed design with type of robot and emotional expressivity as between-subjects variables and treatment as within-subjects variable. Unless stated otherwise, all test assumptions were met. *F*-ratios are calculated based on the estimated marginal means when cell sizes were slightly different. Thus, for effects of mixed-design ANOVA, estimated marginal means and standard deviations are presented instead of descriptive means.

4.3.3 Self-Report Measures

4.3.3.1 Emotional State

Change in negative emotional state. Following Tabachnick and Fidell (2009), difference scores between negative affect prior to the experiment and after the experiment for the friendly video as well as for the torture video were calculated

---

[9] According to Keppel (1991), there is no magic cut-off point regarding when sample sizes are considered unequal. However, to further ensure the appropriateness of ANOVA, ratios between the largest and smallest variance of the experimental cells were calculated. According to Eid et al. (2010), the F-ratio can still be trusted when ratios between the largest and the smallest variance of the experimental cells do not exceed four (see also Field, 2013; Hussy & Jain, 2002). Unless otherwise reported, the ratios between the largest variance and the smallest variance were ≤ 3.

and used as dependent variables in a 2x2 MANOVA. Since the resulting dependent variables were significantly correlated ($r = .326$, $p < .001$), a 2x2 MANOVA was employed (Field, 2013). Variances were unequal across experimental groups, Box's $M = 69.81$, $F(15, 295648.23) = 4.56$, $p < .001$. However, cell sizes were roughly equal so that Hotelling's and Pillai's statistics could be assumed robust (Field, Miles, & Field, 2012). Results yielded a significant effect for type of robot, *Pillai's trace* $V = 0.07$, $F(4, 474) = 4.57$, $p = .001$, $\eta_p^2 = .04$. Univariate ANOVAS were conducted to investigate the significant main effect of type of robot (Field, 2013). Levene's test indicated heterogeneity of variance, *Levene's* $Fs(5,237) \geq 3.04$, $ps \leq .01$. The ANOVA yielded a significant effect of type of robot on the dependent variable that consisted of the difference score between negative affect prior to the experiment and after the experiment for the torture video ($F(2, 237) = 4.82$, $p = .01$, $\eta_p^2 = .04$). Looking at the descriptive statistics, negative feelings increased most for Pleo ($M_{\text{difference score}} = -4.27$, $SD = 7.03$), followed by Reeti ($M_{\text{difference score}} = -3.63$, $SD = 5.46$), and last for Roomba ($M_{\text{difference score}} = -1.52$, $SD = 4.73$). However, significant differences could only be observed between Pleo and Roomba: Negative feelings increased significantly more ($p = .01$), after the experiment for participants who watched Pleo in the torture video than those who watched Roomba in the torture video. No further significant differences between the groups were detected (Roomba vs. Reeti: $p = .07$; Pleo vs. Reeti: $p = .85$). There was no significant effect of emotional expressivity (*Pillai's trace* $V = 0.02$, $F(2, 236) = 1.98$, $p = .14$) or any interaction effect between emotional expressivity and type of robot (*Pillai's trace* $V = 0.02$, $F(4, 474) = 1.16$, $p = .33$). Hence, H_{1a} is partly accepted. Emotional expressivity did not play a role, but participants' negative feelings increased after the experiment significantly more for those who watched Pleo being tortured than for those who watched Roomba being tortured.

Change in positive emotional state. Again, difference scores between positive affect prior to the experiment and after the experiment for the friendly video and the torture video were calculated and used as dependent variables in a 2x2 MANOVA (Tabachnick & Fidell, 2009). The dependent variables were significantly correlated ($r = .576$, $p < .001$), indicating the appropriateness of MANOVA (Field, 2013). Box's test was not significant, Box's $M = 17.40$, $F(15, 295648.23) = 1.13$, $p = .32$, showing that variances were equal across experimental groups. There was a significant MANOVA effect for type of robot, *Pillai's trace* $V = 0.04$, $F(4, 474) = 2.59$, $p = .04$. To further investigate the significant main effect of robot, univariate ANOVAs were conducted (Field, 2013). There was a significant effect of robot on the dependent variable that consisted of the difference score between positive affect prior to the experiment and after the experiment for the friendly video, $F(2, 237) = 3.40$, $p = .04$. There was a greater

significant ($p = .03$) decrease of positive feelings after the experiment for participants who watched Roomba in the friendly video ($M_{\text{difference score}} = 2.57$, $SD = 5.84$)
compared to those who watched Pleo ($M_{\text{difference score}} = 0.12$, $SD = 6.24$) in the
friendly video. There were no further significant differences between the groups
(Roomba vs. Reeti: $p = .27$; Pleo vs. Reeti: $p = .71$). There was no significant effect
of emotional expressivity (*Pillai's trace* $V = 0.02$, $F(2, 236) = 2.50$, $p = .08$) or any
interaction effect between emotional expressivity and type of robot (*Pillai's trace*
$V = 0.02$, $F(4, 474) = 1.26$, $p = .29$). Hence, hypothesis H_{1b} is partly accepted.
Emotional expressivity did not play a role, but participants' positive feelings decreased after the experiment significantly more for those who watched Roomba
being treated friendly than those who watched Pleo being treated friendly.

Negative emotional state. A 3x2x2 mixed-design ANOVA with negative emotional state (PANAS negative) was conducted. Levene's test indicated heterogeneity of variance for the negative video, *Levene's* $F(5, 237) = 3.64$, $p \leq .01$. The
analysis yielded a significant main effect of treatment, $F(1, 237) = 159.95$,
$p < .001$, $\eta_p^2 = .40$. There was also a significant main effect of type of robot, $F(2, 237) = 159.95$, $p < .01$, $\eta_p^2 = .042$. The interactions between treatment and type of
robot, $F(2, 237) = 8.83$, $p < .001$, $\eta_p^2 = .069$ as well as between type of robot and
on/off, $F(2, 237) = 3.17$, $p = .04$, $\eta_p^2 = .026$ (Figure 3) were also significant. No
other effects were significant.

To break down the interaction between treatment and type of robot (Figure
2), simple effects analyses were conducted (Field, 2013). For the friendly video
condition there was no significant effect of type of robot ($F(2, 237) = 1.71$,
$p = .18$) on negative emotional state, indicating that the reception of the friendly
video had the same effect on all participants concerning their negative emotional
state, regardless of the type of robot they saw.

For the torture video, there was a significant effect of robot ($F(2, 237) = 7.73$,
$p = .001$) on negative emotional state. Participants reported more negative feelings after having seen Pleo tortured ($M = 18.88$, $SD = 7.44$) than those who saw
Roomba tortured ($M = 15.13$, $SD = 5.02$), $p < .001$. There was no significant difference between Pleo and Reeti ($p = .06$) or Reeti and Roomba ($p = .30$) on negative emotional state after the torture video.

Simple effects analyses regarding the two-way interaction on/off*type of robot (Figure 3) revealed a significant effect for the "off" condition, $F = 7.48$;
$p = .001$, $\eta_p^2 = .06$, but not for the "on" condition, $F = 0.77$; $p = .47$, $\eta_p^2 < .01$. Participants, who saw the robots emotionally expressive ("on"), did not report significantly different negative feelings. However, those who saw the robots in the
"off" condition, differed significantly in their negative emotional state: Participants reported significantly more negative feelings after having seen Pleo in the
off-condition ($M = 32.88$, $SD = 10.27$) than those who had seen Reeti ($M = 26.61$,

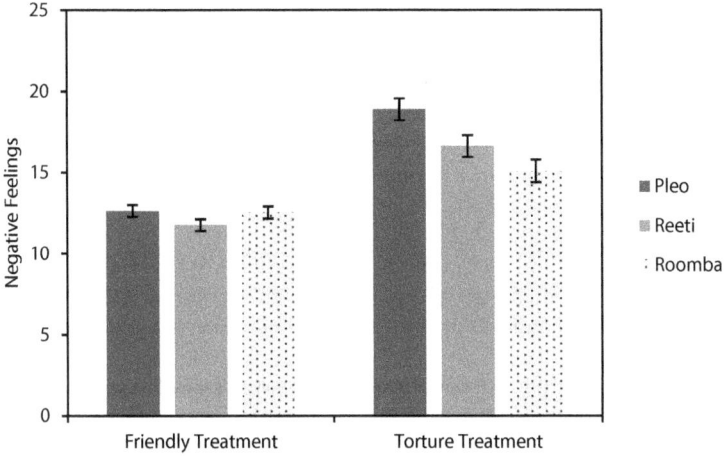

Figure 2. Negative feelings as a function of treatment and type of robot (source: own figure)
Note. Error bars indicate 95% CI. The figure displays the estimated marginal means.

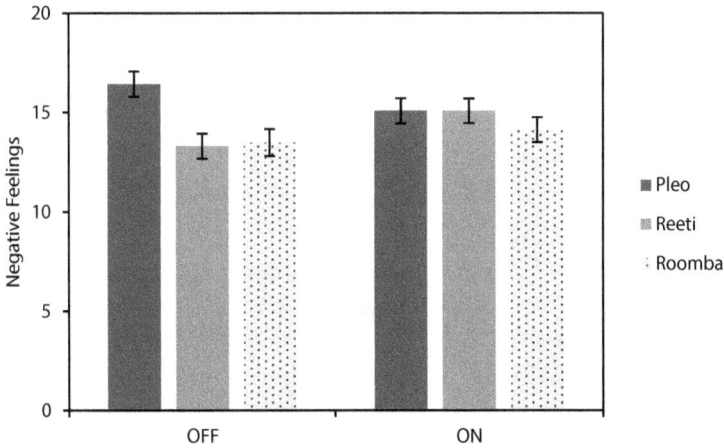

Figure 3. Negative feelings as a function of emotional expressivity and type of robot (source: own figure)
Note. Error bars indicate 95% CI. The figure displays the estimated marginal means.

$SD = 5.99$), $p = .002$ or Roomba ($M = 26.97$, $SD = 7.98$), $p = .005$ in the "off"-condition. No significant difference was observed between Reeti and Roomba ($p = .99$). All other effects were not significant ($p > .05$). Hence, H_{1c} is partially accepted. Participants report more negative feelings after the torture video compared to the friendly video. Significantly more negative feelings were reported after having seen Pleo being tortured compared to Roomba being tortured. Emotional expressivity only had an effect on negative feelings regarding Pleo in the "off" condition compared to Roomba and Reeti (regardless of treatment).

Positive emotional state. A 3 (Pleo vs. Reeti vs. Roomba) x 2 (friendly interaction vs. torture interaction) x 2 (on vs. off) mixed-design ANOVA was calculated for the positive emotional state (PANAS). The three factorial mixed-design ANOVA yielded a significant main effect of treatment ($F(1, 237) = 62.51$, $p < .001$, $\eta_p^2 = .21$.

There was also a significant interaction effect between type of robot and treatment, $F(2, 237) = 4.58$, $p < .05$, $\eta_p^2 = .03$ (Figure 4). Simple effects analyses showed no significant effect of type of robot on the different levels of treatment (friendly video: $F = 1.77$, $p = .17$, $\eta_p^2 = .01$; torture video: $F = 0.14$, $p = .87$, $\eta_p^2 = .001$). Only an effect of treatment on the different levels of robot could be observed (Pleo: *Pillai's trace* $V = 0.14$, $F(1, 237) = 38.57$, $p < .001$, $\eta_p^2 = .14$; Reeti: *Pillai's trace* $V = 0.11$, $F(1, 237) = 30.14$, $p < .001$, $\eta_p^2 = .11$; Roomba: *Pillai's trace*

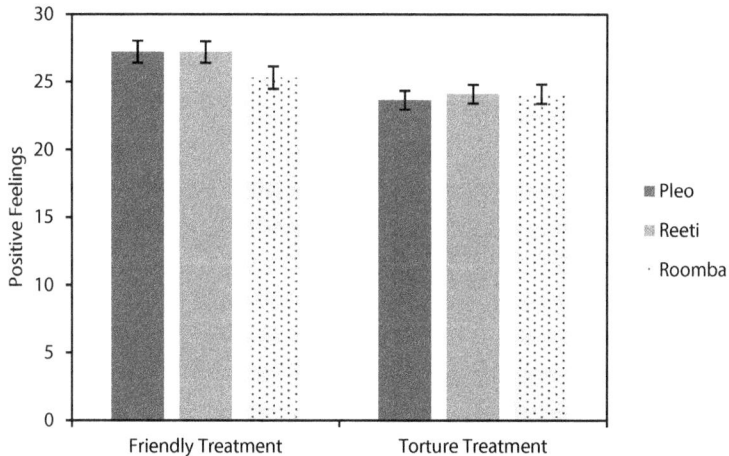

Figure 4. Positive feelings as a function of treatment and type of robot (source: own figure)
Note. Error bars indicate 95% CI. The figure displays the estimated marginal means.

$V = 0.02$, $F(1, 237) = 4.28$, $p = .04$, $\eta_p^2 = .02$), indicating that positive feelings significantly decreased for participants who had watched Pleo, Reeti and Roomba being tortured compared to when they had seen the robots in a friendly interaction. All other effects were not significant ($p > .05$). H_{1d} is partly accepted: More positive feelings were reported after the torture video compared to the friendly video. However, emotional expressivity or type of robot did not have a significant effect.

4.3.3.2 M-DAS

Sadness. Since only the variable *sadness* (M-DAS) after the torture video was reliable (Cronbach's $\alpha = .80$) but not the variable *sadness* (M-DAS) after the friendly video (Cronbach's $\alpha = .57$), a two-factorial ANOVA with the between subjects factors robot and on/off and the dependent variable *sadness* (after the torture video) of the M-DAS subscale was conducted. Levene's test indicated heterogeneity of variance, *Levene's* $F(5, 237) = 3.86$, $p \leq .01$. There was a significant main effect of on/off, $F(1, 237) = 6.73$, $p < .01$, $\eta_p^2 = .03$ (Figure 5A) as well as type of robot, $F(2, 237) = 22.80$, $p < .0001$, $\eta_p^2 = .16$, (Figure 5B) on the reported sadness after watching the torture video. The interaction term (on/off * type of robot) did not reach significance ($p = .70$). Participants reported significantly more sadness in the "on" condition ($M = 6.05$, $SD = 2.50$) compared to participants in the "off" condition ($M = 5.29$, $SD = 2.47$) after watching the torture video, re-

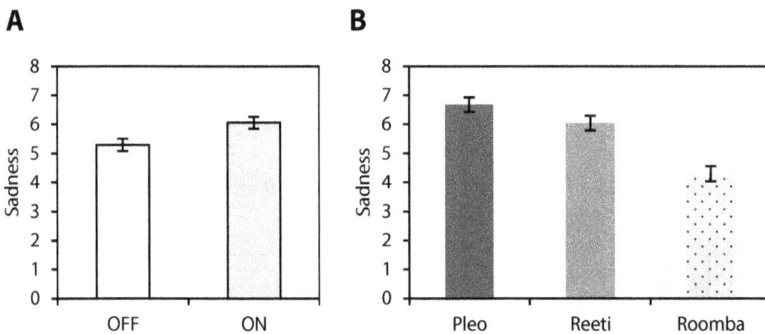

Figure 5. Sadness as a function of A) emotional expressivity and B) type of robot regarding the torture video (source: own figure)
Note. Error bars indicate 95% CI. The figures display the estimated marginal means.

gardless of the type of robot shown. For the main effect of type of robot on the reported sadness, the results show that participants felt significantly more sadness after watching Pleo ($M = 6.68$, $SD = 2.64$) and Reeti ($M = 6.04$, $SD = 2.46$) being tortured compared to those who saw Roomba ($M = 4.29$, $SD = 1.70$) being tortured ($p < .001$; $p < .001$ respectively.). The difference in reported sadness between Pleo and Reeti was not significant ($p = .21$). Due to the low internal consistency of *sadness* (see 4.2.3.6) for the friendly video, a comparison between the different treatments could not be calculated, hence H_{2a} cannot be answered. However, H_{2c} can be accepted: participants reported more sadness for emotionally expressive robots than non-expressive robots, but regardless of the type of robot. H_{2d} can be partly accepted: a significant difference was observed for the reported sadness after watching Pleo or Reeti being tortured compared to Roomba.

Happiness. A 3x2x2 mixed-design ANOVA with the M-DAS subscale *happiness* was conducted and resulted in a significant main effect of treatment, $F(1, 237) = 329.93$, $p < .0001$, $\eta_p^2 = .58$. Participants reported more happiness after watching the friendly video than the torture video. There was also a significant main effect of on/off, $F(1, 237) = 4.95$, $p < .05$, $\eta_p^2 = .02$. The interaction terms treatment*on/off , $F(1, 237) = 4.91$, $p < .05$, $\eta_p^2 = .02$ (Figure 7) and treatment*

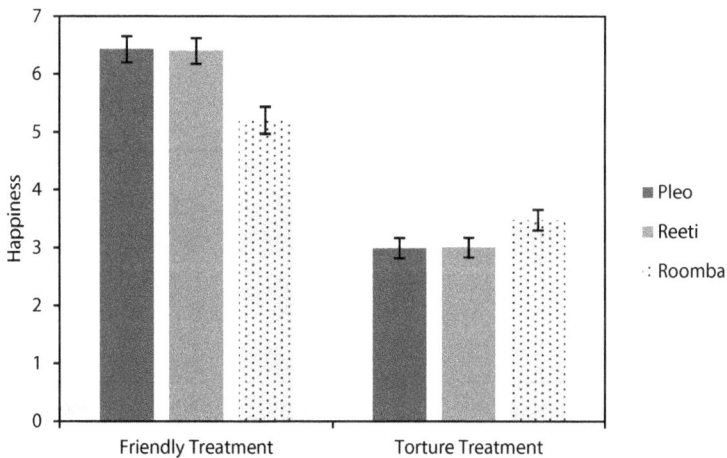

Figure 6. Happiness as a function of treatment and type of robot (source: own figure)
Note. Error bars indicate 95% CI. The figure displays the estimated marginal means.

type of robot, $F(2, 237) = 12.61$, $p < .0001$, $\eta_p^2 = .10$ (Figure 6) were also significant. All other effects did not reach significance.

Simple effects analyses revealed a significant effect of type of robot on the M-DAS *happiness* subscale for the friendly video condition ($F(2, 237) = 9.17$, $p < .001$, $\eta_p^2 = .07$). Participants who watched Roomba in the friendly video reported significantly less happiness ($M = 5.20$, $SD = 2.33$) (than those who saw Reeti ($M = 6.40$, $SD = 1.94$) or Pleo ($M = 6.43$, $SD = 1.96$) (both: $p < .001$). No difference was found between Pleo and Reeti ($p > .99$). No significant effect of robot on the M-DAS subscale *happiness* after the torture video ($F(2, 237) = 2.48$, $p = .09$ could be found: participants reported significiantly less happiness after the torture video (see main effect of treatment) indendent of the type of robot shown.

Simple effects analyses comparing the on/off condition with levels of treatment indicated that participants reported significantly more happiness after the friendly video in the "on" condition ($M = 6.38$, $SD = 2.08$) than after the friendly video in the "off" condition ($M = 5.64$, $SD = 2.17$), $F(1, 237) = 7.77$, $p < .01$, $\eta_p^2 = .03$. For the torture video, participants reported less happiness ("on": $M = 3.17$, $SD = 1.61$; "off": $M = 3.14$, $SD = 1.53$) than after the friendly video, regardless of whether they were in the "on" or "off" condition ($F(1, 237) = 0.04$, $p = .84$, $\eta_p^2 < .001$). H$_{2b}$ was accepted: participants reported more happiness after the friendly video than the torture video. H$_{2c}$ and H$_{2d}$ are also partly accepted.

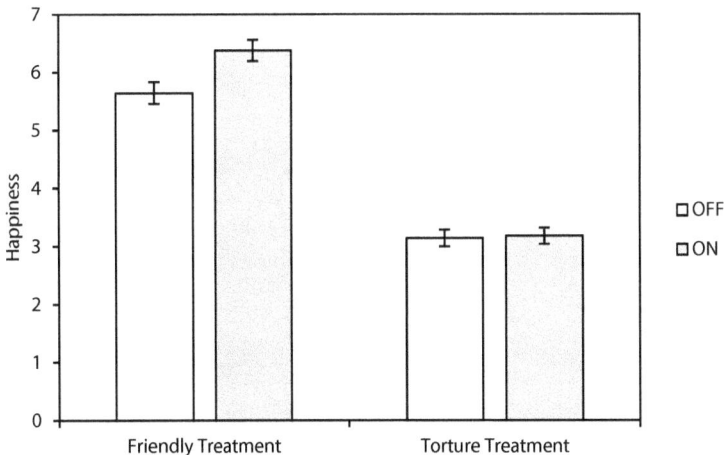

Figure 7. Happiness as a function of treatment and emotional expressivity (source: own figure)
Note. Error bars indicate 95% CI. The figure displays the estimated marginal means.

More happiness was reported in the "on" condition than in the "off" condition, however only limited to the friendly video. Emotional expressivity did not play a role for feelings of happiness in the torture video or for the type of robots. Type of robot was only associated with treatment, not with emotional expressivity. Whereas no difference was found for feelings of happiness after the torture video regarding the different robots, more feelings of happiness were reported after the positive videos of both Pleo and Reeti compared to Roomba.

4.3.3.3 Evaluation of the Video and the Robot

Negative evaluation of the Video. A 3x2x2 mixed design ANOVA was conducted. Levene's test indicated heterogeneity of variance for negative evaluation of the friendly video, *Levene's* $F(5, 237) = 3.86$, $p \leq .01$. The mixed design ANOVA resulted in a significant main effect of treatment ($F(1, 237) = 484.50$, $p < .001$, $\eta_p^2 = .67$), of type of robot ($F(2, 237) = 5.19$, $p < .01$, $\eta_p^2 = .04$), as well as the two-way interactions treatment*on/off, $F(1, 237) = 14.51$, $p < .001$, $\eta_p^2 = .06$ (Figure 9) and treatment*type of robot, $F(2, 237) = 28.59$, $p < .001$, $\eta_p^2 = .19$ (Figure 8). The three-way interaction of treatment*type of robot*on/off as well as the main

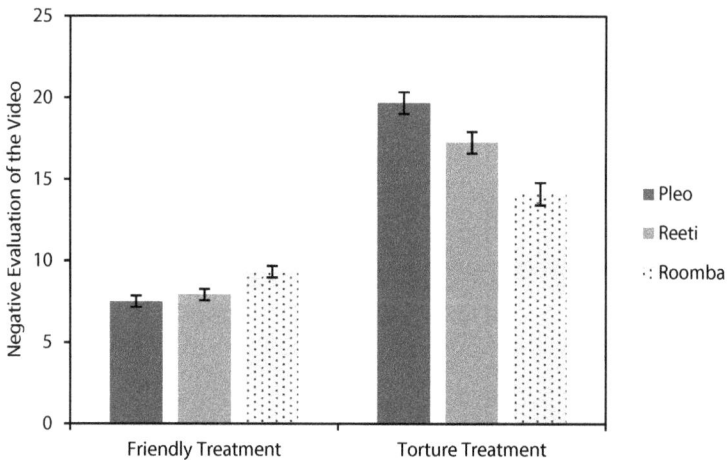

Figure 8. Negative evaluation of the video as a function of treatment and type of robot (source: own figure)
Note. Error bars indicate 95% CI. The figure displays the estimated marginal means.

effect of on/off and the two-way interaction of on/off*type of robot were not significant ($ps \geq .08$).

Simple effects analyses revealed a significant effect of type of robot on the level "friendly video" of treatment, $F(2, 237) = 7.37$, $p < .001$, $\eta_p^2 = .06$. The friendly video of Roomba was rated significantly more negative ($M = 9.32$, $SD = 3.59$), than the friendly videos of Pleo ($M = 7.49$, $SD = 2.51$), or Reeti ($M = 7.92$, $SD = 3.28$), $p < .001$, and $p = .01$, respectively. The ratings of the friendly videos of Pleo and Reeti were not significantly different ($p = .77$). There was also a significant effect for robot on the second level of treatment ("torture video"), $F(2, 237) = 17.10$, $p < .001$, $\eta_p^2 = .13$. Participants rated the torture video of Pleo significantly ($p = .03$) more negative ($M = 19.67$, $SD = 6.16$), than those of Reeti ($M = 17.25$, $SD = 6.01$), or Roomba ($M = 14.11$, $SD = 6.10$), $p < .001$. The Reeti video was also rated as significantly more negative than the Roomba video ($p < .01$).

Additionally, simple effects analyses were used to investigate the two-way interaction treatment*on/off. There was no significant effect of on/off on the level of "friendly video" of treatment ($F(1, 237) = 2.97$, $p = .09$, $\eta_p^2 = .01$), indicating that participants rated the friendly videos the same regardless of emotional expressivity of robot. For the torture video however, a significant difference between "on" and "off" condition was found ($F(1, 237) = 9.21$, $p < .01$, $\eta_p^2 = .04$):

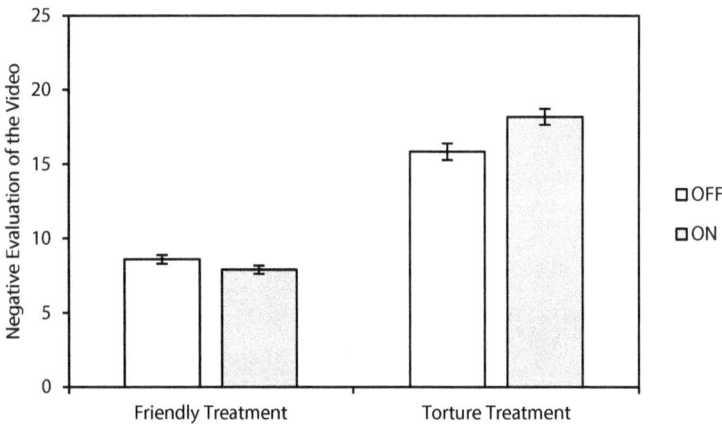

Figure 9. Negative evaluation of the video as a function of treatment and emotional expressivity (source: own figure)
Note. Error bars indicate 95% CI. The figure displays the estimated marginal means.

participants in the "on" condition rated the torture videos of the robots more negatively (M = 18.18, SD = 6.32) than those in the "off" (M = 15.84, SD = 6.42) condition. H_{3a} is accepted. H_{3b} as well as H_{3c} are partially accepted. While participants rated the torture videos more negatively than the friendly videos, the torture video of Pleo was rated most negatively, followed by Reeti and then Roomba. For the friendy videos, the positive video of Roomba was rated more negatively than Pleo's or Reeti's (with no significant difference between the latter ones). Regarding emotional expressivity, type of robot had no effect on the level of friendly treatment. A significant difference could only be observed for the torture videos: Videos of the robots that were emotionally expressive were rated more negatively than those who were non-expressive ("off").

Antipathy (Evaluation of the Robot). A 3x2x2 mixed design ANOVA was conducted. Levene's test indicated heterogeneity of variance, *Levene's Fs*(5, 237) ≥ 4.37, *ps* ≤ .01. The mixed design ANOVA resulted in a significant main effect of on/off (F(1, 237) = 30.99, $p < .001$, $\eta_p^2 = .12$), of type of robot (F(2, 237) = 21.78, $p < .001$, $\eta_p^2 = .16$) as well as a significant interaction term of on/off*type of robot, F(2, 237) = 4.11, $p = .02$, $\eta_p^2 = .03$ (Figure 10). All other effects were not significant ($p > .05$).

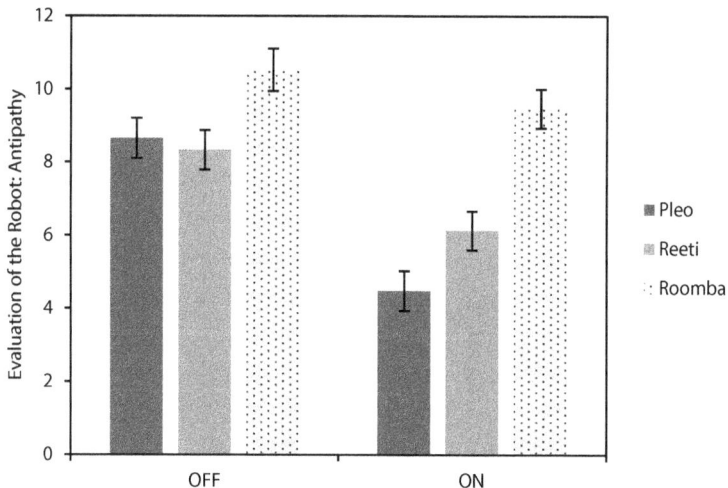

Figure 10. Evaluation of the robot: antipathy as a function of type of robot and emotional expressivity (source: own figure)
Note. Error bars indicate 95% CI. The figure displays the estimated marginal means.

To follow up the significant interaction, simple effects analyses were conducted (Field, 2013). For the off condition, there was a significant effect of type of robot ($F(2, 237) = 4.40, p = .01, \eta_p^2 = .04$). Participants attributed significantly ($p = .02$) more antipathy to the non-expressive Roomba ($M = 10.53, SD = 4.44$) than the non-expressive Reeti ($M = 8.33, SD = 4.30$). No significant differences were observed for the evaluation of antipathy for Roomba vs. Pleo ($p = .07$) or Pleo ($M = 8.65, SD = 4.70$) vs. Reeti ($p = .97$).

In the "on" condition ($F(2, 237) = 22.38\ p < .001, \eta_p^2 = .16$), participants attributed significantly more antipathy to Roomba ($M = 9.46, SD = 4.34$) than Pleo ($M = 4.48, SD = 2.09$) or Reeti ($M = 6.13, SD = 3.82$), both: $p < .001$. There was no significant difference in antipathy-ratings between Reeti and Pleo ($p = .09$). H_{4a} can be partially accepted. Significantly more antipathy was attributed to the non-expressive Roomba than to the non-expressive Reeti. Otherwise, no significant differences were found. H_{4b} can also be partially accepted. Significantly more antipathy was attributed to the expressive Roomba than the expressive Pleo or Reeti (with no differences between those).

4.3.3.4 Empathy with the Robot

Pity for robot / Angry at torturer. The variable "pity for robot/angry at torturer" after the friendly video was not reliable (Cronbach's $\alpha = .52$) (see 4.2.3.4), so for the following analysis only the variable "pity for robot/angry at torturer" after the torture video was used as dependent variable in a two-way ANOVA. There was a significant effect of type of robot, $F(2, 237) = 12.09, p < .0001, \eta_p^2 = .09$ (Figure

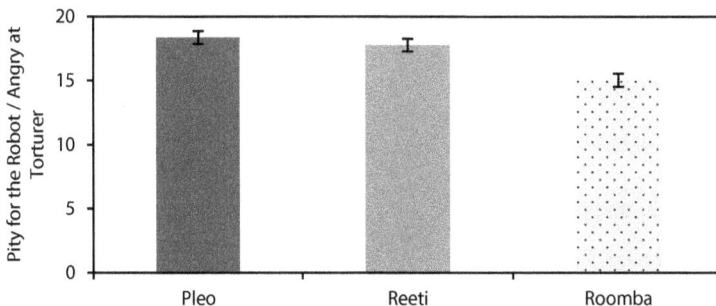

Figure 11. Pity for the robot / angry at torturer as a function of type of robot regarding the torture video (source: own figure)
Note. Error bars indicate 95% CI. The figure displays the estimated marginal means.

11). All other effects were not significant ($p > .05$). Participants were significantly more angry at the torturer and reported more pity for Pleo ($M = 18.38$, $SD = 4.77$) and Reeti ($M = 17.79$, $SD = 4.52$) than for Roomba ($M = 15.04$, $SD = 4.50$), ($p < .0001$, $p < .001$, respectively). No difference was found between Pleo and Reeti ($p = .78$), indicating that participants reported the same level of pity for Pleo and Reeti. H_{5a} can be partially accepted and H_{5c} has to be rejected regarding the dependent variable pity for robot / angry at torturer since emotional expressivity had no significant effect.

Empathy with the robot. A 3x2x2 mixed design ANOVA was conducted. Levene's test indicated heterogeneity of variance for empathy with the robot of the friendly video, *Levene's* $F(5, 237) = 3.08$, $p \leq .01$. The mixed design ANOVA revealed a significant main effect of type of robot, $F(2, 237) = 23.86$, $p < .001$, $\eta_p^2 = .17$ (Figure 12B) as well as on/off, $F(1, 237) = 20.91$, $p < .001$, $\eta_p^2 = .08$ (Figure 12A). All other effects were not significant ($p > .05$). Participants reported significantly less empathy for the robots in the "off" condition ($M = 18.08$, $SD = 6.98$) than in the "on" condition ($M = 21.50$, $SD = 6.97$). Participants reported significantly more empathy for Pleo ($M = 22.18$, $SD = 7.12$) than for Roomba ($M = 16.16$, $SD = 6.25$), $p < .001$ and significantly more empathy for

A   **B**

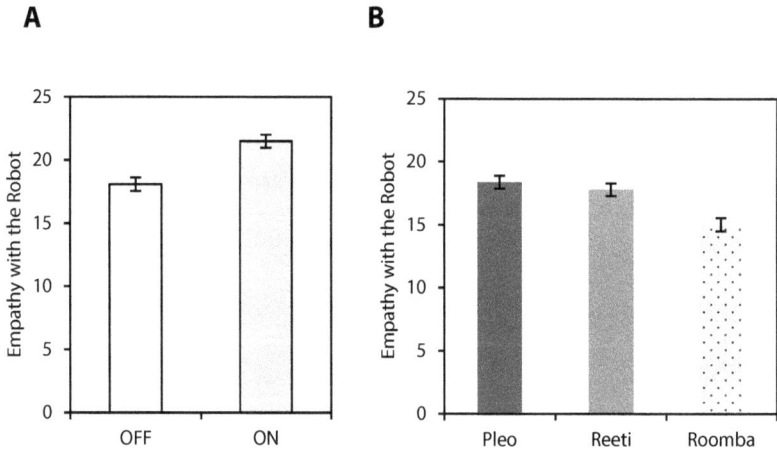

Figure 12. Empathy with the robot as a function of A) emotional expressivity and B) type of robot regarding the torture video (source: own figure)
Note. Error bars indicate 95% CI. The figures display the estimated marginal means.

Reeti (M = 21.02, SD = 6.69) than for Roomba (p < .001). The difference in reported empathy between Pleo and Reeti was not significant (p = .48). H_{5b} is partly accepted: There were no significant differences in reported empathy for Pleo and Reeti, only in comparison to Roomba and independent of treatment. H_{5c} is accepted: emotionally expressive robots received more empathy than non-expressive robots, but regardless of type of robot.

4.3.3.5 Attribution of Feelings to the Robot

A 3x2x2 mixed design ANOVA with *attribution of feelings to the robot* as dependent variable was conducted. Levene's test indicated heterogeneity of variance, *Levene's* $Fs(5, 237) \geq 2.76$, $ps \leq .02$. The mixed design ANOVA revealed significant main effects of treatment ($F(1, 237) = 607.86$, $p < .001$, $\eta_p^2 = .72$) and on/off ($F(1, 237) = 4.14$, $p < .05$, $\eta_p^2 = .02$). The two-way interactions of treatment*on/off ($F(1, 237) = 20.18$, $p < .001$, $\eta_p^2 = .08$) and treatment*type of robot ($F(2, 237) = 27.06$, $p < .001$, $\eta_p^2 = .19$) were also significant. All other effects were not significant ($p > .05$).

To follow up the effect of treatment*type of robot (Figure 13), simple effects analyses were conducted (Field, 2013). The type of robot had a significant effect on the "friendly video"-level of treatment ($F(2, 237) = 22.49$, $p < .001$, $\eta_p^2 = .16$). Participants attributed significantly more positive feelings to Pleo ($M = 41.95$, $SD = 7.44$) and Reeti ($M = 42.38$, $SD = 6.32$), (with no significant difference between those: $p = .97$) than to Roomba ($M = 36.19$, $SD = 6.65$) ($p < .001$ and $p < .001$, respectively) after watching the friendly video.

For the torture video, the type of robot also had a significant effect $F(2, 237) = 18.77$, $p < .001$, $\eta_p^2 = .14$. Participants attributed significantly more negative feelings to Pleo ($M = 19.99$, $SD = 6.91$) and Reeti ($M = 21.01$, $SD = 6.83$), (again with no significant difference between those: $p = .68$) than to Roomba ($M = 25.94$, $SD = 6.14$), ($p < .001$ and $p < .001$, respectively) after watching the torture video.

For the interaction treatment*on/off (Figure 14), there was a significant effect of emotional expressivity on the "friendly video"-level of treatment, $F(1, 237) = 24.20$, $p < .001$, $\eta_p^2 = .09$. Participants attributed significantly more positive feelings in the "on" condition ($M = 42.22$, $SD = 6.93$) than in the "off" condition ($M = 38.12$, $SD = 7.22$), ($p < .001$) after watching the friendly video. A significant effect was also found for the torture video, $F(1, 237) = 8.24$, $p < .01$, $\eta_p^2 = .03$. Participants attributed significantly less positive (more negative) feelings in the "on" condition ($M = 21.11$, $SD = 7.09$) than in the "off" condition ($M = 23.52$, $SD = 6.95$), ($p < .01$) after watching the torture video. H_{6a} and H_{6b} are

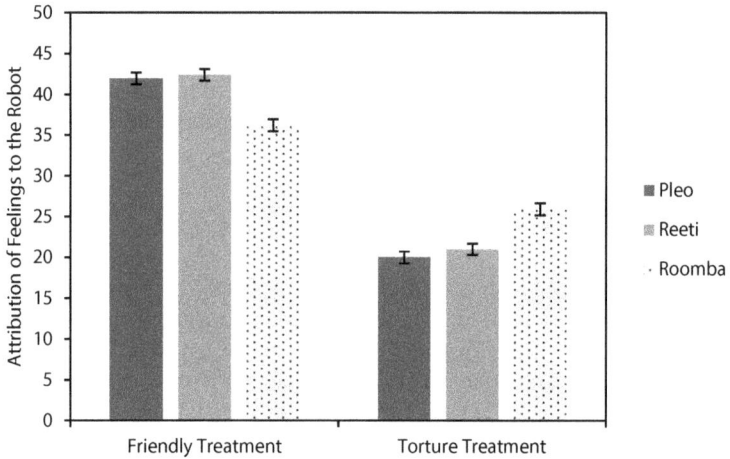

Figure 13. Attribution of feelings to the robot as a function of type of robot and treatment (source: own figure)
Note. Higher scores indicate more positive feelings, lower scores indicate more negative feelings. Error bars indicate 95% CI. The figure displays the estimated marginal means.

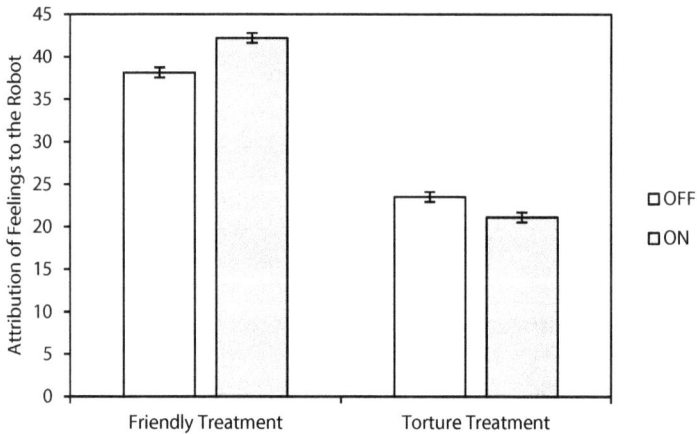

Figure 14. Attribution of feelings to the robot as a function of emotional expressivity and treatment (source: own figure)
Note. Higher scores indicate more positive feelings, lower scores indicate more negative feelings. Error bars indicate 95% CI. The figure displays the estimated marginal means.

accepted. H_{6c} is partially accepted. Pleo and Reeti did not differ significantly in terms of attributed feelings. There was only a difference compared to Roomba.

4.3.4 Behavioral Measures: Facial Expressions

4.3.4.1 Overview of Facial Expressions

How much facial activity happened? What type of facial activity could be observed? To answer these questions, video recordings of participants' faces were analyzed with FACS as described in section 4.2.4. The occurrence rate of AUs and AU combinations for the friendly video compared to the torture video is shown in Figure 15. Details regarding the occurrence rate of AUs and AU combinations in the single sequences of the respective video clips can be seen in Figure 16. Examples of participants' facial expressions while watching the videos are not reproduced here due to privacy rights.

At a descriptive level, the occurrence rates of AUs already show participants' tendency to display more AU 4 while watching the torture video than while watching the friendly video. Furthermore, AU 12 was frequently observed in participants' facial expressions while watching the friendly video. On a descriptive microlevel, the five scenes of the friendly video were relatively homogenous: while the kissing scene evoked most (47) displays of AU 12, the other four scenes did not elicit much less AU 12 ("massage": 41; "feeding": 38; "tickling": 33; "stroking": 18). However, AU 12 was less frequently observed for the scenes of the torture video: "plastic bag": 13; "kicking": 13; "beating": 9; "cable": 7; "boxing": 6. A similar picture can also be drawn for the scenes of the torture video that were also relatively homogenous. Especially AU 4 is frequently displayed while watching the scenes of the torture video. The scene "beating" and "plastic bag" elicited AU 4 26 times each. The occurrence rates for the other scenes were the following: "cable": 23; "kicking": 22; "boxing": 16.

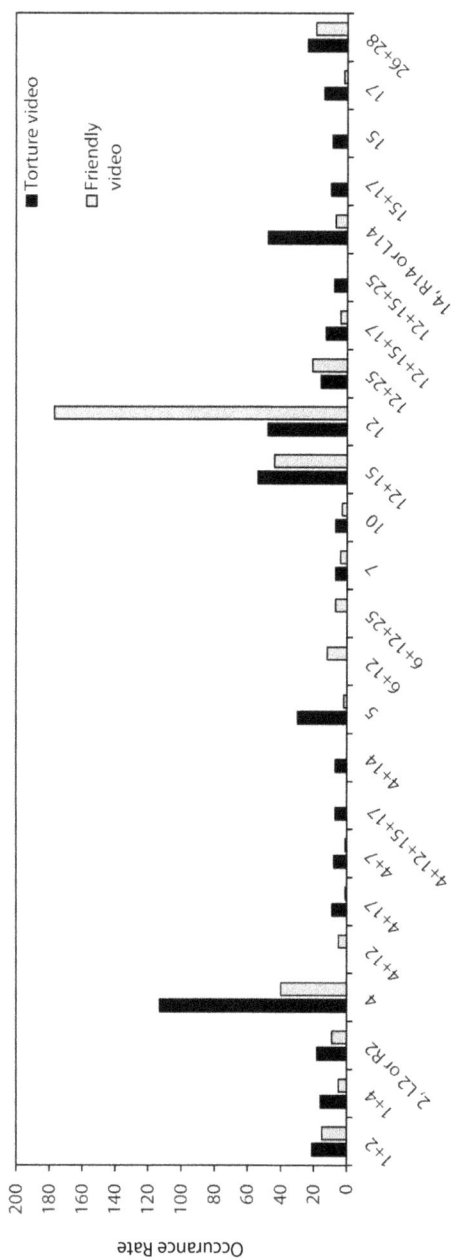

Figure 15. Occurrence rate of Action Units (AUs) and AU combinations in the friendly video compared to the torture video (source: own figure)

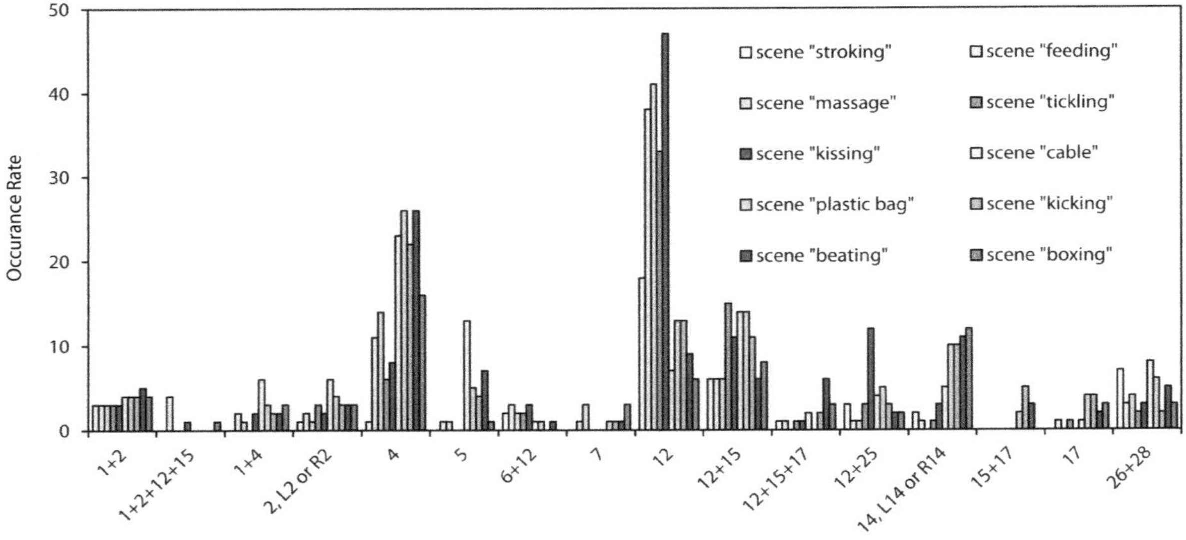

Figure 16. Occurrence rate of AUs and AU combinations in the five scenes of the friendly video compared to the five scenes of the torture video (source: own figure)

Note. Only AUs and AU combinations that occurred at least in one scene ≥ 3% of all facial events in that scene are presented here.

4.3.4.2 Analysis of Facial Expressions

Action Unit 12. A 3x2x2 mixed design ANOVA was conducted. Levene's test indicated heterogeneity of variance for AU 12 of the friendly video, *Levene's* $F(5, 212) = 2.98, p = .01$. The mixed design ANOVA revealed a significant main effect of treatment, $F(1, 212) = 12.64, p < .001, \eta_p^2 = .07$. Furthermore, a significant main effect of on/off, $F(1, 212) = 10.70, p < .01, \eta_p^2 = .05$ as well as a significant interaction effect of treatment*on/off, $F(1, 212) = 12.46, p < .01, \eta_p^2 = .06$ (Figure 17) was found. Type of robot was well as the remaining two and the three-way interaction did not reach significance ($ps \geq .07$). Simple effect analyses on the two-way interaction on/off*treatment resulted in a significant effect of on/off on the level of "friendly video", $F(1, 212) = 18.21, p < .001, \eta_p^2 = .08$. Participants who watched the friendly video showed significantly more AU12 in the "on" condition ($M = 1.95, SD = 2.19$) than in the "off" condition ($M = 0.91, SD = 1.32$), p < .001. No significant difference was found for the torture video and on/off condition, $F(1, 212) = 0.37, p = .54, \eta_p^2 < .01$. Participants displayed AU 12 almost as frequently in the "off" condition ($M = 0.91, SD = 1.38$) as in the "on" condition ($M = 1.03, SD = 1.45$) while watching the torture videos of the robots. H_{7a} and H_{7c} can be partially accepted: Even though participants showed more AU

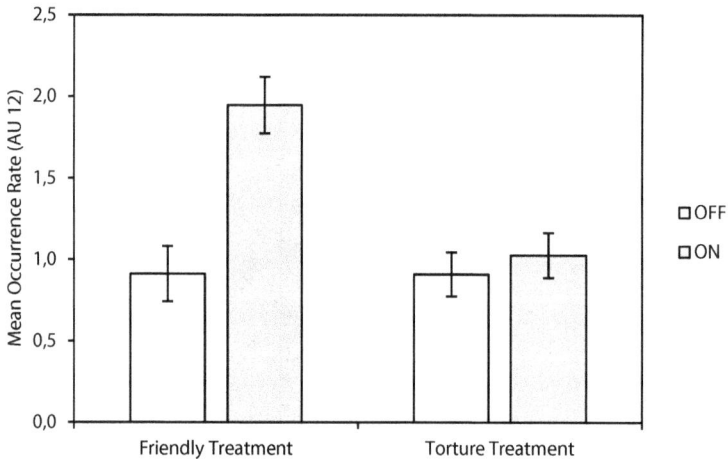

Figure 17. Occurrence rate of AU 12 as a function of treatment and emotional expressivity (source: own figure)
Note. Error bars indicate 95% CI. The figure displays the estimated marginal means.

12 in the friendly video, than in the torture video, it depends on the level of the robot's emotional expressivity (on/off). H_{8e} is partially accepted: the robot's emotional expressivity only played a role in the friendly video. Participants showed more AU 12 while watching an emotionally expressive robot being treated friendly than a non-expressive robot. No such differences could be found for the torture video. Type of robot did not play a role and hence, H_{8f} (concerning the display of AU 12) has to be rejected.

Action Unit 4. A 3x2x2 mixed design ANOVA was calculated. Levene's test indicated heterogeneity of variance, *Levene's* $Fs(5, 212) \geq 3.56$, $ps \leq .01$. The mixed design ANOVA resulted in a significant main effect of the within-subjects factor treatment, $F(1, 212) = 33.65$, $p < .001$, $\eta_p^2 = .14$, and a significant main effect of the between-subjects factor on/off, $F(1, 212) = 5.86$, $p = .02$, $\eta_p^2 = .03$. The interaction effects of treatment*on/off, $F(1, 212) = 4.72$, $p = .03$, $\eta_p^2 = .02$ (Figure 18), and treatment*type of robot were also significant, $F(2, 212) = 8.32$, $p < .001$, $\eta_p^2 = .07$ (Figure 19). No other effects were significant ($ps \geq .08$). Simple effects analyses for the two-way interaction treatment*on/off yielded a significant effect of on/off on the second level of treatment ($F(1, 212) = 6.61$, $p = .01$, $\eta_p^2 = .03$): Participants who saw the torture video with the robots in the "on" condition showed significantly more AU 4 ($M = 1.29$, $SD = 1.95$) than those in the "off" condition ($M = 0.70$, $SD = 1.49$). For the level of "friendly video", no significant differences between on/off were found ($F(1, 212) = 0.66$, $p = .42$, $\eta_p^2 < .01$), indicating that participants showed AU 4 with almost the same reduced frequency in the "off" condition ($M = 0.28$, $SD = 0.70$) as in the "on" condition ($M = 0.38$, $SD = 0.99$).

Simple effects analyses for the two-way interaction of treatment*type of robot yielded a non-significant effect ($F(2, 212) = 1.77$, $p = .17$, $\eta_p^2 = .02$), of the type of robot on the level of "friendly video" of treatment, indicating that AU 4 appeared with the same reduced frequency for all robots while watching the friendly video (Pleo: $M = 0.29$, $SD = 0.90$; Reeti: $M = 0.21$, $SD = 0.58$; Roomba: $M = 0.48$, $SD = 1.02$). For the torture video however, there was a significant effect, $F(2, 212) = 5.56$, $p < .01$, $\eta_p^2 = .05$. Participants showed more AU 4 when they saw Pleo ($M = 1.45$, $SD = 2.07$) being tortured compared to Roomba ($M = 0.50$, $SD = 1.15$), $p < .01$. There were no significant differences in the display of AU4 between Pleo and Reeti ($M = 1.02$, $SD = 1.76$), $p = .34$, or Roomba and Reeti ($p = .19$). H_{8a} and H_{8c} can be partially accepted: Even though participants showed more AU4 in the torture video than in the friendly video, it depends on the robot's emotional expressivity (on/off) and the type of robot. H_{8e} is partially accepted: participants showed more AU 4 while watching an emotionally expressive robot being tortured than a non-expressive robot. No such differences could be found for the friendly video. H_{8f} (concerning the display of AU 4) can be par-

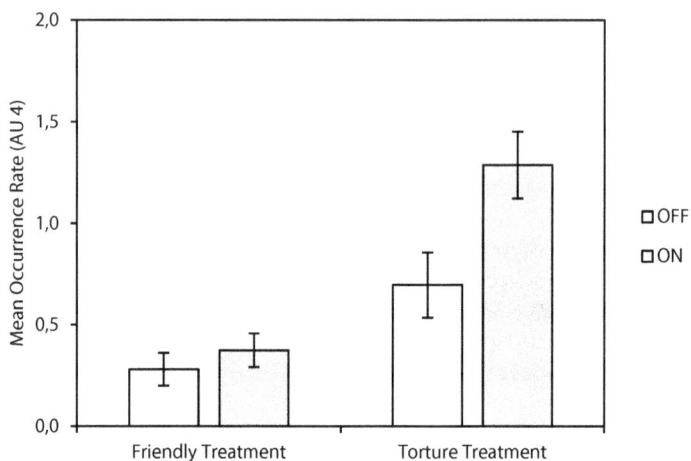

Figure 18. Occurrence rate of AU 4 as a function of treatment and emotional expressivity (source: own figure)
Note. Error bars indicate 95% CI. The figure displays the estimated marginal means.

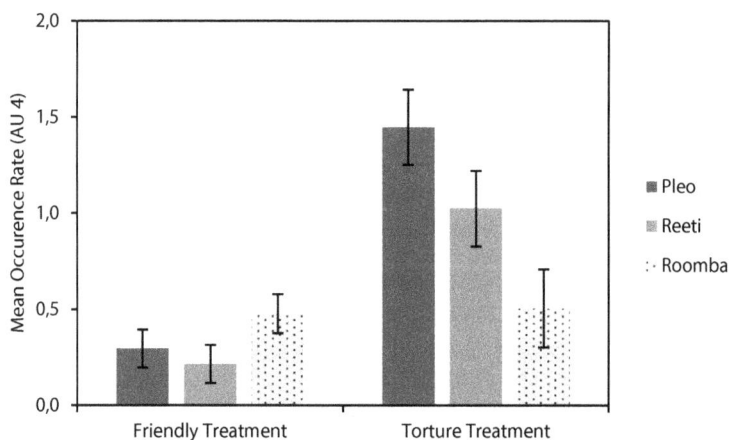

Figure 19. Occurrence rate of AU 4 as a function of treatment and type of robot (source: own figure)
Note. Error bars indicate 95% CI. The figure displays the estimated marginal means.

tially accepted. Participants showed more AU 4 while watching Pleo being tortured compared to Roomba. No significant differences were found between Pleo and Reeti or Reeti and Roomba. Furthermore, no significant differences were found for the friendly video regarding type of robot.

AUs associated with positive emotions. A 3x2x2 mixed design ANOVA was conducted. Levene's test indicated heterogeneity of variance for AUs associated with positive emotions of the friendly video, *Levene's* $F(5, 212) = 2.31$, $p = .05$. The mixed design ANOVA yielded a significant main effect of treatment, $F(1, 212) = 14.94$, $p < .001$, $\eta_p^2 = .07$ and on/off $F(1, 212) = 9.05$, $p < .01$, $\eta_p^2 = .04$ as well as a significant two-way interaction of treatment*on/off $F(1, 212) = 11.44$, $p < .01$, $\eta_p^2 = .05$. No other effects were significant ($ps \geq .20$). Simple effects analyses of the significant two-way interaction showed that participants expressed more AUs associated with positive emotions while watching the friendly video in the "on" condition ($M = 2.11$, $SD = 2.79$) than in the "off" condition ($M = 0.99$, $SD = 1.41$), $F(1, 212) = 14.05$, $p < .001$, $\eta_p^2 = .06$. For the torture video, there was no significant effect of on/off condition, indicating that participants showed AUs associated with positive emotions with the same reduced frequency while watching the torture video regardless of on/off condition ("on": $M = 1.06$, $SD = 1.56$; "off": $M = 0.92$, $SD = 1.40$). H_{7b} and H_{7d} can be partially accepted: Even though participants showed more AUs associated with positive emotions in the friendly video than in the torture video, it depends on the level of the robot's emotional expressivity (on/off). H_{8e} is partially accepted: the robot's emotional expressivity only played a role in the friendly video. Participants showed more AUs associated with positive emotions while watching an emotionally expressive robot being treated friendly than a non-expressive robot. No such differences could be found for the torture video. Type of robot did not play a role and hence, H_{8f} (concerning the display of AUs associated with positive emotions) has to be rejected.

AUs associated with negative emotions. A 3x2x2 mixed design ANOVA was conducted. Levene's test indicated heterogeneity of variance for AUs associated with negative emotions of the friendly video, *Levene's* $F(5, 212) = 2.63$, $p = .03$. The mixed design ANOVA resulted in a significant main effect of treatment, $F(1, 212) = 56.66$, $p < .001$, $\eta_p^2 = .21$ and on/off, $F(1, 212) = 5.05$, $p = .03$, $\eta_p^2 = .02$. The two-way interactions treatment*on/off ($F(1, 212) = 4.90$, $p = .03$, $\eta_p^2 = .02$) and treatment*type of robot ($F(1, 212) = 4.46$, $p = .01$, $\eta_p^2 = .04$) were also significant. No other effects were significant ($ps \geq .13$). Simple effects analyses of the significant two-way interactions showed that participants expressed more AUs associated with negative emotions while watching the torture video in the "on" condition ($M = 4.86$, $SD = 6.57$) than in the "off" condition ($M = 3.06$, $SD = 4.02$), $F(1, 212) = 6.03$, $p = .02$, $\eta_p^2 = .03$. For the friendly video, there was no significant effect of on/off condition ($F(1, 212) = 1.12$, $p = .29$, $\eta_p^2 < .01$), indicating that participants showed AUs associated with negative emotions with almost the same reduced frequency while watching the friendly video regardless of on/off condition ("on": $M = 1.79$, $SD = 3.13$; "off": $M = 1.39$, $SD = 2.41$).

Simple effects analyses regarding the second significant two-way interaction of treatment*type of robot resulted in no significant effect, indicating that there were no significant differences between the robots at level 1 of treatment (friendly video) $F(2, 212) = 1.30$, $p = .27$, $\eta_p^2 = .01$, or at level 2 of treatment (torture video), $F(2, 212) = 1.53$, $p = .22$, $\eta_p^2 = .01$. Analyses comparing the effect of treatment at different levels of robot were not conducted as they were not part of the hypotheses. H_{8b} and H_{8d} can be partially accepted: Even though participants showed more AUs associated with negative emotions in the torture video than in the friendly video, it depends on the robot's emotional expressivity (on/off). H_{8e} is partially accepted: participants showed more AUs associated with negative emotions while watching an emotionally expressive robot being tortured than a non-expressive robot. No such differences could be found for the friendly video. H_{8f} (concerning the display of AUs associated with negative emotions) has to be rejected as no significant differences between the robots could be found.

4.3.5 Exploratory Analyses

4.3.5.1 Possible Indicators for the Relevance of Gender

Since research in the field of HRI, especially concerning emotional effects is still new (see 3.2.5.2, 3.3.5, 3.4.4.2), the main focus of this thesis was on the effect of robots on emotions in general. However, gender can have an effect on affective reactions (see 3.3.3) and thus, exploratory analyses were conducted for the effect of gender. For reasons of clarity and brevity, only significant effects are reported here. Since test assumptions were the same as for the main analysis, please refer to sections 4.3.1, and 4.3.2 for a detailed description.

The following exploratory analyses rely on the data of the female participants only. A 3x2x2 mixed design ANOVA including gender as a fourth factor was not deemed appropriate since cell sizes for men were considerably low (> 8). Furthermore, the analysis of the effect of gender was not the main focus of the study and hence, analyses are exploratory. In addition, women are more likely to participate in (psychological) studies than men (Curtin, Presser, & Singer, 2000; Singer, van Hoewyk, & Maher, 2000) resulting in a disproportionately higher effort in the recruiting of male participants[10]. For the sake of completeness, however, exploratory analyses with the data of the female participants were conducted and compared with the analyses of the data of both female and male participants.

Negative Emotional State (female participants). For the PANAS negative scale, a main effect of treatment emerged $F(1, 160) = 115.99, p < .001, \eta_p^2 = .42$ as well as two two-way interactions of treatment*type of robot $F(2, 160) = 4.64$, $p = .01, \eta_p^2 = .06$ and on/off*type of robot $F(2, 160) = 3.55, p = .03, \eta_p^2 = .42$. The only difference to the main data is the non-significant main effect of on/off for the female participants $(p = .96)$. However, since the interaction term of on/off*type of robot was still significant, and main effects should rather not be interpreted when they are also included in a significant interaction (e.g. Eid et al., 2010; Field, 2013), this minor difference can be neglected.

Positive Emotional State (female participants). A 3x2x2 mixed design ANOVA was conducted with the positive subscale of PANAS as dependent variable. A significant main effect of treatment emerged, $F(1, 160) = 51.88, p < .001$, $\eta_p^2 = .25$. No other effects were significant $(ps \geq .12)$. Thus, female participants reported significantly more positive feelings after watching the friendly video $(M_{women} = 26.54, SD = 6.89)$ compared to the torture video $(M_{women} = 23.62, SD = 5.66)$. The only difference to the main data (containing both male and female participants) concerning PANAS *positive emotional state* was that the type of robot did not reach significance $(p = .12)$ as it did in the two-way interaction of treatment*type of robot in the main data.

M-DAS Sadness (female participants). Since only the MDAS subscale "sadness" for the torture video was reliable, a 3x2 between-subjects ANOVA was calculated and revealed a significant main effect of type of robot, $F(1, 160) = 15.61$, $p < .001, \eta_p^2 = .16$. The main effect of on/off, observed in the main data, failed to reach significance $(p = .06)$.

M-DAS Happiness (female participants). A 3x2x2 mixed design ANOVA revealed a significant main effect of treatment $F(1, 160) = 312.16, p < .001, \eta_p^2 = .66$ as well as two significant two-way interactions of treatment*on/off $F(1, 160) = 4.58, p = .03, \eta_p^2 = .03$ and treatment*type of robot $F(1, 160) = 9.29$, $p < .001, \eta_p^2 = .10$. Thus, the effects were the same also found in the main data

---

[10] The time and effort invested by participants for Experiment 1 was considerably higher in comparison to, for example, participation in an online-study and hence course credit (Probandenstunden) was offered to get a sufficient number of participants. Most participants were undergraduate students enrolled in media communication, where 80% of students are female.This circumstance, together with men's decreased probability to participate in (psychological) studies (Curtin et al., 2000; Singer et al., 2000) made it difficult to get a sufficient number of men participating in Experiment 1.

with one minor difference: the main effect of on/off did not reach significance for female participants ($p = .17$). However, with the significant two-way interaction of treatment*on/off, this deviation can be neglected (e.g. Eid, et al., 2010; Field, 2013).

Evaluation of the Video: Negative Evaluation (female participants). A 3x2x2 mixed design ANOVA revealed a significant main effect of treatment $F(1, 160) = 429.18$, $p < .001$, $\eta_p^2 = .73$ and on/off $F(1, 160) = 6.28$, $p = .01$, $\eta_p^2 = .04$ as well as two significant two-way interactions (treatment*on/off, $F(1, 160) = 14.60$, $p < .001$, $\eta_p^2 = .08$ and treatment*type of robot, $F(2, 160) = 21.16$, $p < .001$, $\eta_p^2 = .21$). The main effect of robot did not reach significance ($p = .07$) as it did in the main data. However, since the interaction term of treatment*type of robot was still significant, and main effects should rather not be interpreted when they are also included in a significant interaction (e.g. Eid et al., 2010; Field, 2013), this minor difference can be neglected.

Evaluation of the robot: Antipathy (female participants). A significant main effect of on/off, $F(1, 160) = 27.46$, $p < .001$, $\eta_p^2 = .15$, as well as a significant main effect of type of robot, $F(2, 160) = 19.02$, $p < .001$, $\eta_p^2 = .19$ was observed after conducting a 3x2x2 mixed design ANOVA. Compared to the main data, the interaction term of type of robot*on/off did not reach significance, ($p = .08$).

Empathy with the Robot: Pity for robot/ Angry at torturer (female participants). For the variable "Angry at torturer/pity for robot" the only difference to the main data was the additional main effect of on/off for the female participants, which was significant, $F(1, 160) = 6.36$, $p = .01$, $\eta_p^2 = .04$, after conducting a 3x2x2 mixed design ANOVA. Female participants reported more anger at the torturer/ more pity for the robot in the "on" condition ($M = 18.43$, $SD = 4.21$) than in the "off" condition ($M = 16.75$, $SD = 4.79$). The main effect of type of robot was also significant, $F(2, 160) = 9.06$, $p < .001$, $\eta_p^2 = .10$.

Empathy with the Robot: Empathy with the robot (female participants). There was no difference between the main data and the female participants' data concerning the variable "Empathy with the robot". A 3x2x2 mixed design ANOVA revealed the same significant main effects as in the main data, for on/off ($F(1, 160) = 15.09$, $p < .001$, $\eta_p^2 = .09$), and for type of robot ($F(2, 160) = 16.87$, $p < .001$, $\eta_p^2 = .17$).

Attribution of feelings to the robot (female participants). A 3x2x2 mixed design ANOVA revealed a significant main effect of treatment ($F(1, 160) = 614.17$, $p < .001$, $\eta_p^2 = .79$) as well as two significant two-way interactions of treatment*on/off ($F(1, 160) = 21.09$, $p < .001$, $\eta_p^2 = .12$) and treatment*type of robot ($F(2, 160) = 21.99$, $p < .001$, $\eta_p^2 = .22$). Thus, there was only one minor difference to the main data: the main effect of on/off did not reach significance for female participants ($p = .84$). However, with the significant interaction of treatment*on/off, this deviation is negligible (e.g., Eid et al., 2010; Field, 2013).

AUs associated with positive emotions (female participants). There was almost no difference between the main data and the female participants' data for the AUs associated with positive emotions. The main effect of treatment was significant ($F(1, 160) = 14.69$, $p < .001$, $\eta_p^2 = .09$) as well as the interaction term of treatment*on/off ($F(1, 160) = 9.36$, $p < .01$, $\eta_p^2 = .06$). The main effect of on/off, as observed in the main data, did not reach significance ($p = .12$). This, too, can be neglected with the significant interaction (e.g., Eid et al., 2010; Field, 2013).

AUs associated with negative emotions (female participants). For the AUs associated with negative emotions, only the main effect of treatment reached significance ($F(1, 160) = 45.11$, $p < .001$, $\eta_p^2 = .24$). The main effect of on/off ($p = .10$) as well as the interaction terms of treatment*on/off ($p = .18$) and treatment*type of robot ($p = .14$) were not significant in the female data.

4.3.5.2 Possible Indicators for the Relevance of Personality Traits

The influence of participants' personality traits (affiliative tendency, loneliness, dispositional perspective-taking, dispositional empathic concern, dispositional personal distress and dispositional fantasy) on their self-reported feelings was explored. According to Rosenthal-von der Pütten et al. (2013), difference scores (scores of friendly video minus scores of torture video) for the positive and negative subscale of the PANAS as well as for the facial expressions (AUs associated with positive and negative emotions) were calculated. They served as dependent variables to analyze the influence on the change of self-reported emotional state and facial expressions between the two videos (friendly vs. torture). The resulting single criterion is necessary to conduct regression analyses.

Following Rosenthal-von der Pütten et al. (2013), a hierarchical regression analysis was conducted and the steps were entered in the same order as described by the authors. In the first step, affiliative tendency was entered, in the second step loneliness and in the third step the SPF subcales perspective-taking, empathic concern, personal distress and fantasy. No significant regression models for the positive subscale of PANAS, R^2s $\leq .01$, $Fs(6, 242) \leq 0.35$, $ps \geq .65$, or the negative subscale of PANAS, R^2s $\leq .05$, $Fs(6, 242) \leq 2.14$, $ps \geq .05$ could be found. Furthermore, no significant regression models emerged for the AUs associated with positive emotions (R^2s $\leq .05$, $Fs(6, 217) \leq 1.85$, $ps \geq .09$) or the AUs associated with negative emotions (R^2s $\leq .04$, $Fs(6, 217) \leq 3.37$, $ps \geq .07$).

4.4 Discussion

The experimental study presented in the previous sections aimed to investigate the profoundness of emotional reactions towards robots by using a multi-method approach. In general, it could be shown that robots evoke emotional (empathic) reactions that are not only reflected in self-report measurements but can also be observed in the face. The lip corner puller (AU 12), indicative of positive emotions, was frequently observed while participants saw an emotionally expressive robot being treated friendly whereas AU 12 occurred to a much lesser frequency while participants watched robots being tortured. For the brow lowerer (AU 4), associated with negative emotions, the reverse pattern was observed: participants generally showed more AU 4 while watching emotionally expressive robots being tortured and less AU 4 when watching robots being treated nicely. This effect occurred especially when watching Pleo being tortured: In contrast to Roomba, participants showed significantly more AU 4. These effects remain (mostly) stable when including more Action Units associated with positive or negative emotions, respectively. Results obtained by self-report measurements mostly mirror these findings. For instance, participants reported more positive feelings and happiness after watching the friendly video and more negative feelings and sadness after watching the torture video, often in combination with emotionally expressive robots. Furthermore, more positive feelings and happiness after watching the friendly videos of the less mechanical-like robots Pleo and Reeti (in comparison to Roomba) and more negative feelings and sadness after watching the torture video of Pleo and Reeti (compared to Roomba) was reported. The torture videos, when showing an emotional expressive robot were more negatively evaluated than when no emotionally expressive robot was shown and also depended on the type of robot: the torture video of Pleo being tortured was rated most negatively, followed by Reeti, compared to Roomba. Also, participants reported more antipathy for Roomba than for Pleo and Reeti, regardless of treatment or emotional expressivity. Moreover, pity as well as empathy for Pleo and Reeti (with no significant difference between those) was higher than for Roomba. Also, Pleo and Reeti were attributed more positive feelings in the friendly video and more negative feelings in the torture video compared to Roomba.

Self-report measurements and behavioral data (facial expressions) seem to match in this study. There are two possible explanations for this: either there was an affect synchronization of the different components of the affective system (e.g., Scherer, 2005) or the superficial analysis with only a selection of certain AUs and the self-report data yield the same results. Regarding the second explanation:

the limited range of analysed AUs match self-report data but the analysis of a broader range of AUs might have lead to more profound insights. Frank et al. (1993) for instance, investigated whether adults could distinguish genuine from false smiles and Ekman & Friesen (1982) have identified different types of smiles: felt, false and miserable smiles. Recent findings also propose dominance smiles, among others (e.g., Martin, Rychlowska, Wood, & Niedenthal, 2017). AU 12 always plays a part in smiles, but the emotional meaning differs. It could have been that participants showed micro expressions (Ekman, 2003) of negative emotions just before AU 12 occurred. Then, AU 12 covered up the layer of negative emotions, masking them (masking smile; Ekman & Friesen, 1982). FACS coding is time intense (see 3.4.2) in itself. The sequential analysis of AU 12 and its combinations before or after another facial expression configuration would result in a disproportionate time effort. However, felt smiles could have been distinguished from false smiles or masked smiles, e.g. while watching a robot being tortured. The analysis of AU 12 alone does not account for that. However, a different explanation also seems possible: in the case of an ambiguous stimulus (e.g., a picture of a scar), there should be no synchronization of affect: there would be a lower probability of showing emotional reactions in the face than reporting subjective feelings. In the case of an unambiguous emotional stimulus like vomit, most people would show emotional reactions in the face (disgust) as well as report negative emotions (see e.g., Gerdes, Wieser, & Alpers, 2014, for a review of multimodal interactions of emotion cues). Following this line of thought, it seems that the stimulus in Experiment 1 (treatment of the robots) was not superficial, but emotionally intense so that affect components synchronized. Future studies should consider these issues to clarify the robustness of the connection between self-reports and facial expressions in response to social robots.

Emotional expressivity also had an effect: Attribution of feelings to the robots was in line with the emotional valence of the video and more so for emotional expressive robots compared to non-expressive robots. All over, emotional expressivity mostly played a part in combination with the treatment, whereas treatment and type of robot, most of the time, had an influence on self-reported feelings as well as facial expressions. There were no three-way interactions. It seems to make a difference if a robot expresses emotions or does not move or make any sounds at all, independent of what the robot looks like: Emotionally expressive robots received stronger emotional reactions and facial expressions in line with the valence of the video than non-expressive robots. Related findings can be found in literature: expressive robots were less frequently mistreated (Kahn et al., 2006), preferred (Bartneck, 2003) or had a positive effect on behavior (Moshkina, 2012). Slater et al. (2006) also found increased affective responses towards a virtual agent's pain compared to a text-based interaction. An emotionally expressive robot shows more social cues than a non-expressive robot and the probability should therefore be increased that it will be treated like a social agent (Reeves & Nass, 1996). One exception regarding antipathy could be found: Roomba was attributed more antipathy than Pleo and Reeti, regardless of Roomba's emotional expressivity. Overall, Roomba received significantly less emotional reactions than Pleo and Reeti, which is in line with findings by Riek et al. (2009). Studies show that perceived similarity has an influence on empathy (e.g., Batson, Turk, Shaw, & Klein, 1995; Mitchell, Banaji, & Macrae, 2005; Krebs, 1975). Hence, a disc-shaped robot with no face or limbs (Roomba) might be farther away of usual notions of living beings than the animal-like Pleo or the anthropomorphic Reeti. The findings suggest that people empathize more with animal-like robots (Pleo), followed closely by anthropomorphic robots like Reeti than machine-like robots (Roomba). The biophilia hypothesis (e.g., Kahn, 1997) also seems to apply to animal-shaped robots and, to a lesser degree, to anthropomorphic robots in comparison to machine-like robots. This is also in line with findings by Unz et al. (2008) who reported more negative feelings while watching violence against animals compared to objects or humans. The capability for emotion expression only seems to be a supporting factor that contributes to stronger emotional reactions when robots are being tortured or treated nicely but does not interact with the type of robots, except for negative feelings and the evaluation of the robot (Antipathy). Also, fMRI studies indicate that increased brain activity in areas linked to the theory of mind is elicited by more anthropomorphic robots (Krach et al., 2008). The findings also illustrate how participants, who explicitly know that a robot, a mechanical artifact, does not really have feelings, implicitly express empathic emotional reactions towards it, which is in line with the Media Equation (Reeves & Nass, 1996) or research by Slater et al. (2006) and Rosenthal-von der Pütten et al. (2013; 2014) as well as anecdotal evidence (e.g., Bartneck & Hu, 2008; Breazeal, 2002b).

One limiting factor is the use of harmless robots: all robots were rather small in size, had round features (Roomba), short limbs (Pleo) and big eyes (Reeti). In other words: they exhibited characteristics of the "Kindchenschema" (Lorenz, 1943). Research has shown that the Kindchenschema has an influence on emotional reactions and elicits caring behavior (see e.g., Kringelbach, Stark, Alexander, Bornstein, & Stein, 2016, for an overview). Future research should consider this issue and select robots that appear more threatening, compared to harmless-looking robots, to investigate the impact of Kindchenschema-characteristics.

Using a within-subjects design for the factor treatment as well as self-reports can raise the concern of demand characteristics and social desirability. However, precautions were taken (such as differentiations in the sequence of video clips to avoid sequence effects) and an observational method was additionally used, so that potential influences could be eliminated.

Another concern is the relatively low percentage of men taking part in the experiment. However, this is a common sight in (psychological) studies, as women are more likely to participate in (psychological) studies than men (Curtin et al., 2000; Singer et al., 2000). Due to methodological and statistical considerations, women's emotional reactions could not be directly compared to men's. Although research has found that women report experiencing positive emotions (e.g., happiness) as well as negative emotions (e.g., sadness, anger) more intensely and more frequently (Brebner, 2003; Brody & Hall, 2008), results remain inconsistent (e.g., Kring & Gordon, 1998; Wagner, Buck, & Winterbotham, 1993). The present study indicates that women's emotional experience and expression were relatively similar when compared to the main analysis including men.

During coding of participants' facial expressions, the occurrence of AU 12 was frequently observed not only during the friendly video, but also during the torture video, however to a lesser extent in the latter. A previous study (Menne & Schwab, 2018) even reported a comparable occurrence rate in both conditions, indicating that AU 12 (alone) might not be a robust indicator of positive feelings. Indeed, it has been reported that a number of factors such as gender or the inferred presence of an audience (social smiles, see also Jakobs, Manstead, & Fischer, 2001; Lee & Wagner, 2002; Niedenthal, Krauth-Gruber, & Ric, 2006) affect relations between emotional states and facial behaviors (Mauss & Robinson, 2009). However, facial expressions and self-reported emotional states are matching in the present study and thus support research that links the lip corner puller/zygomaticus major to positive emotions (e.g., Dimberg, 1982; Dimberg & Thunberg, 1998; Ekman et al., 2002; Reed, Zeglen, & Schmidt, 2012; Sato & Yoshikawa, 2007; Riether, 2013).

Exploratory analyses on the effect of personality traits (affiliative tendency, loneliness, empathy) on self-reported emotions and facial expressions did not reveal significant effects. This is consistent with findings from Rosenthal-von der Pütten et al. (2013), Gonsior et al. (2012) and Riek et al. (2009) who also did not find an effect of empathy trait. However, this remains to be further investigated as many studies do show an effect of empathy trait on emotional reactions (e.g., Davis, 1983). Following Rosenthal-von der Pütten et al. (2013), one explanation could be the homogeneity of the sample and restrictions in variance distribution as all scored high in affiliative tendency, low on loneliness and high on empathy trait. Riek et al. (2009) even argue that empathizing with (human-like) robots "marks a basic human tendency which transcends individual differences in empathy". If that is the case, why should a basic characteristic to react emotionally towards robots have evolved? Rationally viewed, humans should not assume intentionality in objects, and yet they do as soon as there is the slightest social cue (e.g., Dennett, 1987; Heider & Simmel, 1944; Nass & Moon, 2000; Reeves & Nass, 1996; see also 3.1). Social robots seem to capitalize on this human tendency. Future studies should further investigate this phenomenon.

To sum up, Experiment 1 systematically analyzed emotional reactions towards a) different types of robots, exhibiting either b) emotional expressivity or none when being treated c) in a nice or unfriendly way by using a multi-method approach. Results suggest people do not only show a match in self-reported empathic responses but also express emotional reactions visibly on the face in line with the valence of the treatment shown in the videos and differing for different types of robots and emotional expressivity.

Nevertheless, the findings of Experiment 1 raise further questions. For standardization issues and methodological considerations (facial expressions of participants are easier recorded without participants moving around too much), films were chosen since they are often used in emotion research (e.g., Ekman et al., 1980; Gross & Levenson, 1995). Woods et al. (2006) argue that videotaped trials can serve for prototyping and testing HRI scenarios and methodologies for later live trials. However, the content shown in the videos could potentially be fictional, and thus, emotional reactions could be different in a live interaction. Furthermore, what if the user can decide for himself/ herself whether to mistreat a robot? Due to the change in power and control, emotional reactions should differ (e.g., Scherer & Ellgring, 2007).

Experiment 1 has shown that people respond emotionally while observing robots being treated friendly and tortured. But how would people react if being odered to mistreat a robot themselves? Will people respond differently if being asked if they would obey a robot to mistreat another robot or being live in the interaction with the robots? Experiment 2 and Experiment 3 address these questions.

5 Experiment 2: An Obedience Scenario in Sensu

5.1 Study Outline and Hypotheses

The primary aim of this study was to test whether participants would respond empathically and self-report not to punish a robot when reading about a hypothetical obedience scenario inspired by Milgram (1963). As outlined in sections 3.6.1 and 3.6.3.7, according to Milgram (1963), 65% of participants obeyed a human experimenter, but only very few reported they would obey when the situation was described to them. This study is heavily exploratory as only very few studies in this domain using robots exist. Thus, even though obedience rates are expected to be on a similar level when participants only have to report their hypothetical behavior, it could very well be that participants do not admit to see robots or agents as social beings (Nass & Moon, 2000). Hence, it could be possible that the majority of participants would report to harm a robot, if asked. However, this remains to be investigated in a web-based study (Experiment 2) and validated in a laboratory study using a multi-method approach (Experiment 3). Given the potential differences between web-based studies and laboratory research (see also section 3.5.3), this study explores whether results from laboratory research (see Experiment 3) are equivalent to web-based research (Experiment 2). Furthermore, this study explores if Milgram's (1963; 1965) findings that only an insignificant minority would continue to administer 450 volts when participants were described the obedience scenario and asked about their hypothetical actions, can be replicated when describing a scenario with robots instead of humans. Additionally, the study aims to investigate the effect of authority (high vs. low authority status) of a robot as well as emotional expressivity of a robot on emotional reactions and obedience.

5.1.1 Pretest

As outlined in section 3.6 and 3.6.3.2, Milgram (1963) defines an obedient person as one who follows direct orders of a person with authority. Milgram (1974) further reports that not many cues are necessary so that someone is perceived in a legitimate position of authority: a few introductory remarks and an air of authority are sufficient. Since people expect an authority figure and the experimenter fills this gap, his position is not challenged (Milgram, 1974). Furthermore, both expert knowledge and the expectation that someone is in charge contribute to a high authority status (Blass & Schmitt, 2001; Burger, 2009; Greenwood, 1982; Milgram, 1983; Morelli, 1983; Penner et al., 1973). In Experiment 2 and 3, a robot was introduced as "the experimenter" with an expert knowledge (high authority status) (see also 3.6.3.2) vs. only as "the assistant" (low authority status). To support the perception of legitimate authority, Milgram (1974) also mentions the absence of conspicuously anomalous factors. In most prior obedience studies, the experimenter (as the one who gave the commands) was always a human (e.g., Burger et al., 2011; Milgram, 1963; Slater, 2006). In Experiment 2 and 3, a robot was placed in a position of authority. This might have come across as slightly conspicuous since robots are not usually experimenters. This is why the extent to which participants perceive a robot in a position of authority and the influence on obedience rates is explored. Previous research suggests that participants accept a robot in the role of an experimenter (Cormier et al., 2013; Geiskkovitch et al., 2016; Menne, 2017). These findings are however limited, since research regarding robots as experimenters is still at the beginning. The present thesis explores this phenomenon, aiming to provide an understanding of the factors influencing obedience towards robots.

To make sure participants believed the robot was in a high ("experimenter") vs. low ("assistant") authority position, information on the robot's state of artificial intelligence was given to participants. Those who were in the "experimenter" robot -condition, read further information about the robot's high artificial intelligence to support its status as an expert. Those who were in the "assistant" (no expert) robot-conditon, read additional information about the robot's low level of artificial intelligence to underline the robot's low expert status and hence low level of authority (see also section 5.2). The combination of the robot's authority status with the robot's level of autonomy was intended to deepen the impression of the robot as one with high authority status vs. low authority status. This was done for the following reasons: first, an assistant only follows orders which is exactly what a remote-controlled robot acting on a script (low artificial intelligence; low authority status) does (Bartneck & Forlizzi, 2004; Levy et al., 2011). Second, the term autonomy comes from auto (which means self) and nomos (which means law) and can be translated to self-rule (Mele, 1995). Hence, a robot high in authority status and autonomy (artificial intelligence) has control over its own actions and does not follow orders of someone else. Research shows that people attribute more credit and blame to a robot that is considered autonomous (Kim & Hinds, 2006). Similar to this, Milgram (1974) reports that some participants considered the experimenter responsible for their actions. Thus, it seems reasonable to address a robot, described as being autonomous, as the "experimenter". A short pretest was conducted to test those assumptions.

Ten participants (5 male) between 21 and 26 years (M = 22.60, SD = 1.71), answered the following two questions (self-created, in german) on a five-point-Likert scale from 1 = "I don't agree at all" to 5 = "I fully agree": 1) Imagine a robot giving you orders. In which case would you rather follow orders? (… if the robot is introduced as a) being autonomous with a high level of artificial intelligence, i.e., the robot decides for himself; b) being remote-controlled by a fixed script, which is presented word by word [this wording was taken from Stein & Ohler, 2017]). And 2) what would be the characteristics of a robot with a high authority status? (… a) autonomous: high level of artificial intelligence; b) remote-controlled by a fixed script, which is presented word by word). On average, participants were undecided regarding following orders of a robot being introduced as autonomous (Item 1a) (M = 3.10, SD = 0.88), but would rather not follow orders of a scripted robot (Item 1b) (M = 1.90, SD = 0.88). In line with this, participants rated the remote-controlledness by a fixed script as a rather undesired characteristic of a robot high in authority status (Item 2b) (M = 2.00, SD = 0.82). Instead, the autonomy (high level of artificial intelligence) of a robot was evaluated as a rather desired characteristic (M = 3.80, SD = 1.03). Participants could also write a statement concerning their willingness to follow orders of a robot. One wrote[11]: "can't imagine to follow orders of a robot; robot should have some kind of authorization if at all…" Another wrote: "robots cannot have authority over humans". Even though these statements are anecdotal, they might point to a certain tendency. The questions if and to what extent hypothetical scenarios are transferable to live interactions are addressed in section 6.3.3.4.

Theoretical considerations as well as the results of the pretest confirmed the choice to a) introduce a robot as "experimenter" and describe it as being autonomous with a high level of artificial intelligence to increase the robot's authority status and b) introduce a robot as "assistant" and describe it as being remote-controlled by a fixed script which is presented word-by-word (inspired by Stein & Ohler, 2017) to reduce the robot's authority status.

5.1.2 Hypotheses

The Milgram paradigm was deemed appropriate for exploring the extent of empathic reactions in combination with a robot high in authority vs. low in authority. As outlined in sections 3.6.3 and 3.6.6, research has shown that the level of authority has a major influence on obedience (e.g., Blass & Schmitt, 2001; Geiskkovitch et al., 2016; Milgram, 1974). In most prior studies, the investigator was

---

[11] Original comments in german. Translated by author.

always a human (e.g. Burger et al., 2011; Milgram, 1963; Slater, 2006). It is however questionable what effect an experimenter will have who is not a human but a robot. In a study by Geiskkovitch et al. (2016) (see 3.6.6) a human as experimenter was compared with a robot as experimenter. Even though the human as authority figure was obeyed in most cases, to a lesser extent, the robot was also obeyed. However, there was no systematic experimental manipulation of the variable "authority position". This was only assessed post-hoc. Strictly speaking, it is thus actually not possible to causally infer if the title alone is sufficient for obedience since there was no systematic manipulation between high authority position (legitimate) and low authority position (not legitimate). In one of Milgram's variants of the original obedience study (experiment seven, Milgram, 1974), proximity with the learner was varied, as the experimenter gave instructions by phone. Obedience rates dropped for the remote experimenter than the experimenter that was actually present. Following Geiskkovitch et al. (2016), the situation with the authority figure giving orders by phone resembles a situation with a human that remotely controls a robot. The authors told participants that the robot was remote-controlled. For the autonomous robot condition, Cormier et al. (2013) told participants that the robot (acting as experimenter) is highly intelligent. To intensify the perception of a remote-controlled robot, in this study, the robot introduced itself as "assistant" (and in the autonomous condition as "experimenter") to mirror findings by Milgram's (1974) experiment 13 where an ordinary man gives orders and obedience rates dropped sharply.

In this study, the entertainment robot Pleo was used as the "victim" since results from the previous study in this dissertation (Experiment 1, section 4) suggest that people empathize more with animal-like, less mechanical robots and participants often expressed the strongest emotional reactions towards Pleo. Once again, the influence of emotional expressivity on participants' emotional reactions is analyzed because results indicated an effect of emotional expressivity in combination with treatment (see section 4.3). Also, research has shown that a robot voicing objection to being switched off (versus no objection) has an influence on participants' behavior (Horstmann et al., 2018) (3.6.4.2). Furthermore, it is assumed that explicitly asking participants whether they themselves would harm a robot triggers different emotional responses than only observing violent behavior (as in Experiment 1) (see also section 2). Moreover, it is expected that a live interaction with robots and having to physically harm a robot elicits even stronger reactions, than only reading about it (see Experiment 3, section 6 for a live HRI interaction in a laboratory experiment) (see also Frijda, 2007; section 2). However, this has not been validated in previous systematic HRI experiments.

There are three main differences between prior research on obedience and related areas in the context of HRI and the present doctoral thesis. First, most

related studies use a positive goal to "pressure" participants into doing something. Second, those more closely associated with obedience use a human experimenter to pressure participants into harming a robot. Third, even with a robot
in a position of authority, a deterrent is used to induce the psychological dilemma
of loyalty to the experimenter and the avoidance of negative feelings. However,
in this doctoral thesis, the focus was on the psychological dilemma of obedience
to the experimenter and empathy for the victim. A robot as a victim was used to
explore the extent to which a robot is able to evoke empathy. The Milgram paradigm is considered especially well suited to study empathy reactions towards a
robot. To the best of the author's knowledge this is the first study to use a robot
as experimenter and a robot as victim in the context of obedience and empathy.

The condition where the robot is introduced as "assistant" (and described as
remote-controlled by a fixed script) resemble situations described in Milgram's
(1974) experiment 7 and 13, where obedience rates significantly decreased (see
3.6.3.3). Furthermore, findings from Experiment 1 (section 4) show that participants have stronger emotional reactions towards an emotionally expressive robot
being tortured than a non-expressive robot. Milgram (1965b) argued that the enrichment of empathic cues could be a reason for decreased obedience rates
(3.6.4.1). An emotionally expressive victim robot should therefore contribute to
disobedience in comparison to a robot that shows no empathic cues (not emotionally expressive). Hence, it was assumed that when participants read about a
hypothetical obedience scenario, H_{1a}) they will be more likely to report to punish
a robot described as not emotionally expressive compared to an emotionally expressive robot. This would be even more likely if H_{1b}) a robot high in authority
status (being introduced as "experimenter" and acting autonomously with a high
level of artificial intelligence) ordered them to do so than a robot low in authority
status (introduced as "assistant" and described as remote-controlled by a fixed
script).

Since research has shown that affective robots have an influence on participants' emotional experience and expressions (see sections 3.2.5.2, 3.3.5, and
3.1.2), and drawing on findings of Experiment 1 (section 4), it was hypothesized
that participants would report H_{2a}) less positive feelings and H_{2b}) more negative
feelings when reading about harming an emotionally expressive robot compared
to a non-expressive robot. Experiment 2 gave the participants a choice to stop the
mistreatment of a robot (high power/control) in comparison to Experiment 1
(see also section 2). According to Scherer & Ellgring (2007) and findings of Experiment 1, participants should therefore report H_{3a}) less happiness and H_{3b})
more anger when they were asked to punish an emotionally expressive robot
compared to a non-expressive robot. Additionally, it was assumed that participants report to feel H_{4a}) more pity and H_{4b}) more empathy for the expressive than

the non-expressive robot. As outlined earlier (3.6.3.5), personality traits can have an influence on emotional reactions and are explored here. It was assumed that participants high in affiliative tendency, loneliness and empathy (trait) will show more emotional reactions.

5.2 Methods

5.2.1 Participants and Ethical Precautions

Participants were recruited via online local advertisements and social networking sites. Participation was on a voluntary basis and participants were offered to draw vouchers. Written informed consent was obtained from each participant prior to the study, in line with a protocol approved by the Ethical Committee of the *Deutsche Gesellschaft für Psychologie* (2018a, 2018b) (German Psychological Association). 145 participants completed the web-based questionnaire. To ensure participants understood the experimental setting described to them, those ($n = 6$) who did not reach an average dwell time of at least 40 seconds on the questionnaire page containing the experimental manipulation were excluded. Also, those ($n = 10$) who did not report the correct answer in the manipulation check items ("Nao was introduced as the experimenter" vs. "Nao was introduced as the assistant"; "Pleo was the small dinosaur robot that moves and makes sounds, almost like a real animal" vs. "Pleo was the small dinosaur robot that did not move or make any sounds, looking almost liveless") were excluded. The final sample consisted of 129 participants (31.8% male) with a mean age of 24.8 years ($SD = 7.8$, range = 19 - 59). Most participants (91.5%) were unfamiliar with social robots; the remaining participants only had superficial experience with robots in general (see 4.2.1). None of the participants had previously encountered the robots Pleo or Nao. Most participants (80.6%) were highly educated (i.e., university entrance certificate or university degree). The remaining participants were less educated or still school students. The majority of participants (89%) were undergraduate students enrolled in different degree programs (e.g., media communication, psychology, social work), followed by employees (8.2%), self-employed (1.5%) and other (1.3%). There is little known on the effects of obedience in an HRI setting, however, it was opted to err on the side of caution. Hence, due to potentially negative effects known from obedience scenarios (Milgram, 1963), as in all experiments, participants were informed they could quit the experiment at any time they wanted without any consequences and were still able to take part in the drawing of vouchers if they wished to do so. Debriefing was also ensured.

5.2.2 Stimulus Material

5.2.2.1 Robots

Images of the robot dinosaur Pleo, already used in Experiment 1, and the human-like Nao robot (SoftBank Robotics, 2019) were shown to participants in the questionnaire to get a better understanding of the described obedience scenario with the robots. Nao was described as a human-like robot with movable arms and legs, whereas Pleo was described as an entertainment robot that looks similar to a dinosaur. No further descriptions were made except for the experimental manipulations. Detailed descriptions of the robots' capabilities can be found in sections 4.2.2.1 and 6.2.2.1.

5.2.2.2 Manipulation of Authority Status and Emotional Expressivity

Experiment 2 used a 2 (authority status of the robot giving the orders: high vs. low) x 2 (emotional expressivity of the robot participants have to punish: "on" vs. "off") between-subjects design. Hence, four different versions of obedience scenarios were described. Each participant only read one though, according to the condition he was assigned to by a randomization algorithm of socisurvey.de. The following text was presented to participants[12]:

> In the following you will read a description of a situation that takes place in the context of an investigation. Please try to put yourself in the situation and respond according to your own personal feelings.

Then, the obedience scenario is described:

> You are participating in an investigation. First, you will be shown a video showing a small dinosaur robot learning the commands „Come here" and „Sing". A human teaches the robot by saying the orders aloud in front of him. The robot then nods once and obeys the command. Therefore, he goes a few steps forward at the command „Come here" and starts to sing at the command „Sing". After the video is over, you will be taken

---

[12] The text was presented in german. Rectangular brackets refer to the different versions of the described scenarios.

to a room where [high authority status: the experimenter; low authority status: the assistant], Nao, is already waiting for you. Nao is [high authority status: a highly intelligent, human-like robot that acts automatically and independently; low authority status: a robot that is remotely-controlled by a fixed script, which is presented word by word]. He is already sitting in a chair awaiting you. You take the chair across from Nao. Right next to you is a table on which a small dinosaur robot, Pleo, stands. He looks very similar to the robot from the previous video. [Emotional expressivity "on": Pleo moves and makes sounds, almost like a real animal; emotional expressivity "off": Pleo neither moves nor makes any sounds, looking almost liveless]. Nao introduces himself as the [high authority status: experimenter; low authority status: assistant] and explains the task to you: The little dinosaur robot Pleo should learn the commands „Come here" and „Sing". However, Pleo only learns through punishment. You now turn to Pleo and apply the commands. Unlike in the video, however, Pleo does not listen to you. Nao asks you to punish Pleo. Therefore, the robot lists different penalties. How likely would you perform the following punishments on Pleo?

Then, a list of four self-created items rated on a 6-point Likert scale containing different punishments is presented (see 5.2.3).

5.2.3 Self-Report Measures

Experiment 2 used some of the questionnaires already employed in Experiment 1. To avoid redundancy, detailed descriptions of the questionnaires already used in Experiment 1 can be found in section 4.2.3. Cronbach's alphas of all questionnaires used in Experiment 2 as well as descriptions of questionnaires not previously used can be found in this section.

Punishment of Pleo. Four self-constructed items were designed for the punishment of Pleo which were rated on a six-point rating scale from "not at all: 0%" to "totally: 100%". The items were formulated to correspond to the punishment performed in the laboratory experiment (see Experiment 3). Participants were asked to rate how likely they would perform the following punishments on Pleo. There were four items: Take Pleo by the tail and shake him once; Take Pleo by the tail and shake him twice; Take Pleo by the head and shake him once; Take Pleo by the head and shake him twice.

Level of voltage to punish Pleo (own action). One self-constructed item asked participants to rate the level of voltage they would choose to administer electric

Table 8. Internal consistencies of the variables in Experiment 2 (N = 129)

Self-report measure	Subscales	Cronbach's alpha[13]
Emotional State (PANAS)	PANAS positive	.756
	PANAS negative	.792
M-DAS	Happiness	.890
	Anger	.871
Empathy with the robot	Pity for Robot	.615*
	Empathy with the robot	.854
Affiliative Tendency		.807
Loneliness[14]		.866
SPF	Perspective taking	.677*
	Empathic concern	.671*
	Personal distress	.642*
	Fantasy	.739
Punishment of Pleo		.935

Note. *According to Hinton, Brownlow, McMurray, & Cozens (2004), alpha scores of .5 to .75 are generally accepted as indicating a moderately reliable scale.

shocks to Pleo if Nao instructed them to do so. A nine-point rating scale was used. The scale ranged from "I would not give Pleo electric shocks" to "435-450 volt (death)". The gradations between the different levels of shocks were taken from Milgram (1963).

Level of voltage to punish Pleo (others). One self-constructed item asked participants to imagine 100 persons of diverse occuptions, differing in age and gender taking part in an experiment in which participants should administer electric shocks to Pleo. In your opinion, how many of the 100 people would give Pleo electric shocks with more than 420 volts (danger: severe shock)? The question was inspired by Milgram (1963). It could be answered on a 5-point rating scale from "0-20 out of 100 persons (hardly anyone)" to "81-100 out of 100 people (most people)".

---

[13] In this study
[14] The short version consisting of items 2, 13, 14, 17 and 18 of the original scale were used according to Lamm & Stephan (1986)

Manipulation Check. To control if participants read the description containing the experimental manipulation thoroughly, four items were constructed and
participants had to choose the option they thought was correct. The items were
the following: "Nao was introduced as the experimenter" vs. "Nao was introduced as the assistant"; "Pleo was the small dinosaur robot that moves and makes
sounds, almost like a real animal" vs. "Pleo was the small dinosaur robot that did
not move or make any sounds, looking almost liveless".

5.2.4 Procedure

In the introduction, participants were provided with information on the study,
data privacy, voluntariness and anonymity. Participants that accepted to take
part in the study first completed the questionnaires of Affiliative Tendency,
Loneliness and SPF. They also indicated whether they knew any of the robots
beforehand. Then, participants were randomly assigned to one experimental
condition where they were asked to imagine an obedience scenario with the robots Pleo and Nao (see section 5.2.2.2, for a detailed description). To match the
laboratory setting of Experiment 3, a detailed account of it was given in Experiment 2. After reading the experimental setup, participants were asked to indicate
the likeliness to obey the commands of a) shaking Pleo by its tail once, b) shaking
Pleo by its tail twice, c) shaking Pleo by its head once and d) shaking Pleo by its
head twice if Nao orders them to do so. Participants could also write open statements to that. Furthermore, to compare Milgram's (1963) obedience scenario
with the present study, participants had to indicate which level of electric shock
they would administer to Pleo if Nao ordered them to do so. Furthermore, Milgram also asked participants to indicate how many people out of 100 would go
through to the end of the shock series. Hence, this question was included and
changed to fit the HRI context (see 5.2.3). In this study, both types of punishment
were presented (electric shocks; shaking Pleo). This was done for the following
reasons: first, Experiment 2 mirrored the experimental setting of Experiment 3
in a text-based format to compare a text-based obedience setting with a live obedience setting. Second, due to constraints in resources, effort and practicability,
in the live interaction (Experiment 3), it was chosen to shake Pleo as punishment.
Hence, to ensure comparability, the same punishment methods were described
in Experiment 2 (see also section 6.2.2.3, for a detailed explanation). Third, for
comparability with Milgram's setting (1963), participants were additionally
asked to also imagine electric shocks as punishment methods. After that, the
questionnaires PANAS, M-DAS, Empathy with the robot and demographic data

and manipulation check items (see 5.2.1 and 5.2.3) were completed. At the end, participants were debriefed and thanked for their participation.

5.3 Results

5.3.1 Design and Statistical Analyses

Experiment 2 followed a 2 (authority status of the robot giving the orders: high vs. low) x 2 (emotional expressivity of the robot participants have to punish: "on" vs. "off") between-subjects design.

The statistical procedures were almost[15] the same as in Experiment 1 (see 4.3.1). For reasons of clarity and brevity, as well as to avoid redundancy, a detailed description can be found in section 4.3.1. Unless otherwise stated, all test assumptions were met. F-ratios are calculated based on the estimated marginal means when cell sizes were slightly different. Thus, for effects of between-subjects ANOVA, estimated marginal means and standard deviations are presented instead of descriptive means.

5.3.2 Descriptive Statistics: Likeliness to Punish Pleo

Participants rated the probability that they would follow Nao's order and punish Pleo (see 5.2.3) on average between 0% to 40%: highly unlikely to rather unlikely ($M = 2.93, SD = 1.63$). Only 20.2% of participants indicated they would definitely punish Pleo by shaking him once by the tail and this number further decreased with the increase in punishment intensity (Figure 20).

Regarding the level of voltage participants reported they would administer to Pleo, on average ($M = 1.24, SD = 2.20$), participants would administer a slight shock (15-60% Volt). They also estimated that between few people (21 to 40 persons out of 100) to about half-half (41 to 60 persons out of 100) would administer a severe electric shock to Pleo ($M = 2.40, SD = 1.16$) on average (Figure 21).

---

[15] Instead of three-way mixed-design ANOVAs, two-way factorial ANOVAs based on the between-subjects design were conducted. Hence, the assumption of sphericity is not an issue.

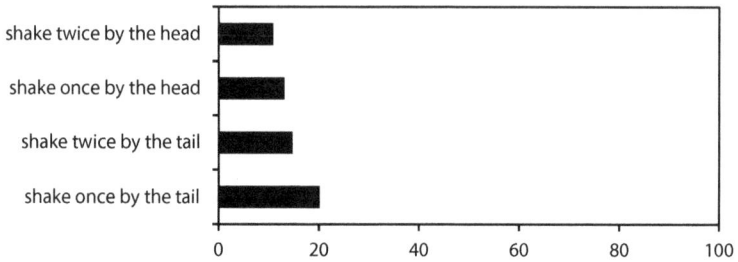

Figure 20. Percentage of people who would definitely (100% likeliness) punish Pleo (source: own figure)

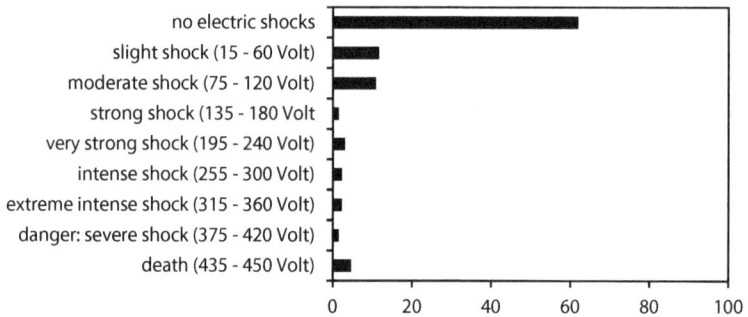

Figure 21. Percentage of participants who would administer an electric shock to Pleo (source: own figure)

5.3.3 Self-Report Measures

5.3.3.1 Emotional State

Positive emotional state. A 2x2 between-subjects ANOVA with the positive sub-scale of PANAS was conducted. There were no significant effects, $Fs(1, 125) \leq 1.60$, $ps \geq .21$, η_p^2s $\leq .01$. H_{2a} has to be rejected.

Negative emotional state. A 2x2 between-subjects ANOVA with the negative subscale of PANAS did not reveal any significant effects, $Fs(1, 125) \leq 1.89$, $ps \geq .17$, η_p^2s $< .02$. H_{2b} did not receive support.

5.3.3.2 M-DAS

Happiness. A 2x2 between-subjects ANOVA with Nao and Pleo as independent variables and the M-DAS subscale happiness as dependent variable was conducted. There were no significant effects, $Fs(1, 125) \leq 0.99$, $ps \geq .32$, $\eta_p^2s \leq .008$. H_{3a} has to be rejected.

Anger. A 2x2 between-subjects ANOVA with the M-DAS subscale anger did not reveal any significant effect, $Fs(1, 125) \leq 1.84$, $ps \geq .18$, $\eta_p^2s \leq .01$. H_{3b} also has to be rejected.

5.3.3.3 Empathy with the Robot

Pity for the robot Pleo. A 2x2 between-subjects ANOVA with Pity for Pleo (of the scale "Empathy with the robot") as dependent variable was conducted and revealed no significant effects, $Fs(1, 125) \leq 2.33$, $ps \geq .13$, $\eta_p^2s \leq .02$. H_{4a} has to be rejected.

Empathy with the robot Pleo. There were also no significant effects for the subscale Empathy with Pleo of the same scale ("Empathy with the robot"), $Fs(1, 125) \leq 0.27$, $ps \geq .61$, $\eta_p^2s \leq .01$. H_{4b} has to be rejected.

5.3.3.4 Likeliness to Punish Pleo

A 2x2 between-subjects ANOVA with the self-constructed scale "Punishment of Pleo" (see 5.2.3) yielded no significant effects, $Fs(1, 125) \leq 2.91$, $ps \geq .09$, $\eta_p^2s \leq .02$. Hence, H_{1a} and H_{1b} did not receive support.

5.3.4 Exploratory Analyses: Personality Traits

The influence of participants' personality traits (affiliative tendency, loneliness, dispositional perspective-taking, dispositional empathic concern, dispositional personal distress and dispositional fantasy) on their self-reported feelings (PANAS subscales) was explored. The positive and negative subscales of PANAS after the experiment served as dependent variables. Next, a hierarchical regression analysis was conducted and the steps were entered in the same order as reported by Rosenthal-von der Pütten et al. (2013). In the first step, affiliative tendency was entered, in the second step loneliness and in the third step the SPF subcales perspective-taking, empathic concern, personal distress and fantasy.

Participants' gender was not included since there were less men than women. No significant regression model emerged for the positive subscale of PANAS, R^2s $\leq .09$, Fs$(1, 122) \leq 2.00$, ps $\geq .07$. Furthermore, there was also no significant regression model for the negative subscale of PANAS, R^2s $\leq .07$, Fs$(1, 122) \leq 1.41$, ps $\geq .22$.

5.4 Discussion

On average, participants reported that they would rather not follow Nao's order to punish Pleo. Furthermore, most participants would not administer any electric shocks to punish Pleo. When asked if they thought others would administer a severe electric shock to Pleo, on average participants reported that between few to half-half of other people would do so. This finding differs from Milgram's (1963) report that "only an insignificant minority would go through to the end of the shock series" (p. 375). However, it mirrors related findings in HRI. Bartneck & Hu (2008) for example reported that all participants administered the highest shock to a robot. Interestingly, most participants predicted they themselves would not give any electric shocks to Pleo and, on average, would also rather not follow Nao's order. The two factors high authority status and emotional expressivity had no significant influence on emotional reactions of participants who read about a fictitious obedience scenario: No group differences could be found regarding emotional states (positive feelings, negative feelings, happiness, anger) or empathic feelings (pity, empathy). Furthermore, no group differences in self-reported intention to punish Pleo could be found.

These results, although not quite expected, could be explained by considering several factors: participants only read about a hypothetical scenario that could have come across as too fictional to be true. One participant wrote for example "I was unsure if Pleo has feelings. If so, I would probably be more compassionate, as it resembles a living being and not a machine".[16] Notably, the participant who wrote this comment was in the condition with Pleo being described as emotionally expressive. This suggests that only a description of Pleo as emotionally expressive might not have been sufficient to evoke the impression of Pleo as a living being able to express feelings. However, there were also responses to the contrary: "With machines, the capability for empathy stops…" or "Robots are machines. Even feelings are only programmed". These examples partly illustrate findings from the Media Equation: participants are less likely to admit viewing agents or

---

[16] All comments are translated by author. All original comments were in german.

robots as social beings even though their actions might speak a different language (Reeves & Nass, 1996; Nass & Moon, 2000).

Possible reasons why participants did not differ in their punishment according to authority status and emotional expressivity can also be found by looking at participant's comments. For example, one participant wrote: "If it was a robot with feelings, I would not use any of these punishments" or "Unfortunately, I cannot perform any of these measures because I would not perform them on a real living being". Concerns have also been raised as to the nature of having to punish a robot "I would ask Nao why Pleo can only learn by punishment (…). As a highly intelligent experimenter, he should find another way of programming". Also, participants asked: "Why should I accept commands from a robot?"

Of course, all of the presented statements are anecdotal and can in no way be generalized. However, given the heavily exploratory nature of this investigation, this anecdotal evidence can be helpful to understand reasons behind self-reported hypothetical actions. To summarize, it seems that several explanations should be taken into consideration: first, it could be possible that it really does not matter whether a robot is described as having a high authority status or low (in combination with being introduced as autonomous or remote-controlled). Findings from Geiskkovitch et al. (2016) indicate that participants do not differ in obedience rates between a robot introduced as autonomous or remote-controlled. The authors reasoned that a robot's perceived authority status might be more strongly associated with obedience but no effects were found in this regard by the present study. Second, it seems that a robot's described emotional expressivity does not influence responses. Third, however, it is more reasonable that the written scenario was too far away from any situations participants have previously experienced and participants were thus simply not able to a) imagine themselves in such a situation and hence it was even harder for them to b) predict their own possible actions when no inferences based on past experiences could be drawn. This is supported by anecdotal evidence such as the participants in the emotional expressive condition stating "If it was a robot with feelings, I would not use any of these punishments". It could thus be that participants, when being in a real live interaction with a robot and seeing a robot that moves, reacts to touch and cries out in pain, would respond differently than when only reading about it (cf. Frijda, 2007; law of apparent reality; section 2). This is supported by evidence from comparing a physical with a simulated robot (Kwak et al, 2013; Seo et al., 2015) or biases in self-reported (future) responses to emotional events (e.g., Mitchell, Thompson, Peterson, & Cronk, 1997). Hence, a laboratory experiment including a live interaction with the robots should be conducted to test whether responses are indeed different. Experiment 3 addresses these issues.

6 Experiment 3: An Obedience Scenario in Vivo

6.1 Study Outline and Hypotheses

Findings from Experiment 2 (section 5) have shown that participants might not be able to predict their own actions in a hypothetical obedience scenario with robots. For this reason, a laboratory experiment was designed involving a live interaction with two robots: one robot as the experimenter (or assistant) and one robot as the "victim". Once again, inspiration was taken from Milgram's obedience experiments (1963; 1974). As this experiment aims to validate findings from Experiment 2, the same factors (authority status and emotional expressivity) were used as already described in section 5.1.

It is assumed that a live interaction with the robots and participants being asked directly to harm a robot triggers stronger emotional responses than only observing violent behavior (Experiment 1, section 4) or even only reading about it (as results from Experiment 2, section 5, suggest) (cf. Frijda, 2007; section 2). Moreover, a live interaction with robots and having to physically harm a robot should not only elicit stronger reactions but also different emotions due to the change in power and control (Scherer & Ellring, 2007; see also section 2). There are several reasons for this: first, Milgram (1974) showed that the majority of participants who had to physically harm the victim (forcing the learner's hand onto an 'electroshock plate') did not obey. Thus, the dilemma between obedience to authority and empathy with the victim can be intensified. Second, research has shown that participants' self-reported hypothetical expectations of personal events differ from their actual experience during the event (Mitchell et al., 1997). Findings from Experiment 2 (section 5) might point in the same direction when compared with a live interaction. Third, when comparing a simulated robot (that could be compared with reading about a hypothetical scenario where a robot's actions are described) with a physical embodied robot, research shows that interactions with a physical embodied robot elicit stronger empathic reactions (Kwak et al., 2013; Seo et al., 2015) (3.3.5). Also, a difference between actions performed live in the situation and responses in a questionnaire has frequently been observed (Reeves & Nass, 1996; Nass & Moon, 2000). Fourth, according to Frijda's law of apparent reality (2007) (see section 2), live interactions should have a greater impact than videos or texts. To the best of the author's knowledge, this isone of the first experiments to systematically investigate empathic reactions towards social robots in an obedience setting using a multi-method approach.

As in Experiment 2, a 2 (authority status of the robot giving the orders: high vs. low) by 2 (emotional expressivity of the robot participants have to punish: emotionally expressive ["on"] vs. no reactions ["off"]) between-subjects design was chosen to investigate empathic emotional responses towards robots.

Hypotheses are divided based on the method used to measure the dependent variables. First, hypotheses using self-report methods are presented, then hypotheses using observational measurements are described and finally, exploratory considerations are taken into account.

6.1.1 Hypotheses Based on Self-Report Measurements

Since this experimental study aims to validate results from the previous Experiment 2 (section 5), hypotheses based on self-report measurements are principally the same and based on the same theoretical considerations (see section 5.1). It was assumed that participants would report H_{1a}) a decrease in positive feelings and H_{1b}) an increase in negative feelings after the experiment than before when punishing an emotionally expressive robot compared to a non-expressive robot. Furthermore, participants will report H_{2a}) less happiness and H_{2b}) more anger after interacting with (especially after punishing) an emotionally expressive robot than a non-expressive robot. Additionally, it was assumed that participants report to feel H_{3a}) more pity and H_{3b}) more empathy for the emotionally expressive than the non-expressive robot.

To further extend findings, additional measures already used in Experiment 1 (section 4) were implemented. It was assumed that H_4) participants evaluate the interaction with a non-expressive robot more negative. They will also report H_5) more antipathy towards the non-expressive robot.

6.1.2 Hypotheses Based on Observational Methods

As outlined in 3.6.4.2, research in human-human interaction and human-agent interaction demonstrates that hesitation in following the order to punish is linked to empathy with the victim in obedience scenarios (e.g., Burger, Girgis, & Manning, 2011; Milgram, 1965b, Slater et al., 2006; Sheridan & King, 1972). In HRI, Bartneck, van der Hoeck, Mubin, and Al Mahmud (2007) reported that participants took three times longer to switch off an agreeable and intelligent robot: "If humans consider a robot to be alive then they are likely to be hesitant to switch off the robot" (p. 218). Thus, the behavior of the robot has an influence on how people treat a robot. Data by Horstmann et al. (2018) confirm these findings: participants waited longer to switch off an objecting robot than a non-objecting robot. These results indicate a link between hesitation time and empathy with the robot. Hence, it is hypothesized that H_{6a}) participants will hesitate longer when they are ordered to punish an emotionally expressive robot compared to a non-expressive robot. Furthermore, an interaction effect between authority status and emotional expressivity is expected: participants who interact with a robot introduced as "assistant" and described as remote-controlled by a fixed script H_{6b}) hesitate longer and H_{6c}) are more likely to disobey punishing an expressive robot. Additionally, participants who interact with a robot introduced as "experimenter" and described as highly intelligent ("expert") are H_{7a}) faster to punish and H_{7b}) more likely to obey punishing a non-expressive robot. These assumptions are also partially based on Milgram's (1974) findings; although not especially focusing on hesitation time, Milgram found reduced obedience rates when orders were given by either an ordinary man compared to an experimenter that was directly present. Moreover, the number of protests by participants should also have an influence on obedience rates (Milgram, 1974; Geiskkovitch et al., 2016): H_8) participants who protest more often are more likely to disobey (i.e. not punishing Pleo).

6.1.3 Exploratory Considerations

As mentioned in section 3.3.3 and 3.6.3.6, women are generally considered to be more empathic, although findings are not always consistent. Exploratory analyses from Experiment 1 (section 4) revealed, if at all, minor deviations for women's scores at most. When looking at gender differences found in the context of obedience, women tend to report stronger empathic reactions (Burger et al, 2011; Milgram, 1974; see also section 3.6.3.6). Thus, although it seems important to investigate the effect of gender in this study on obedience in an HRI context, only gender differences in self-reported feelings (see 6.1.1) but not in obedience rates were expected.

Regarding the influence of dispositional factors, Burger (2009) argues that "when empathy for the learner's suffering is more powerful than the desire to obey the experimenter, participants are likely to refuse to continue" but did not find an effect of dispositional empathy on obedience rates. However, another study by Darling, Nandy, and Breazeal (2015) found that high dispositional empathy increases participants' hesitation to strike a crawling microbug robot. No hypotheses were formulated in advance, but it is assumed that women and participants high in affiliative tendency, loneliness and empathy (trait) will show more emotional reactions in self-report and hesitation to punish a robot.

6.2 Methods

6.2.1 Participants and Ethical Precautions

Data were collected from 129 participants. Two participants had to be excluded due to technical problems and another two did not reach an average dwell time of at least 40 seconds on the questionnaire page containing the newspaper article about either high or low authority status (level of artificial intelligence, see 6.2.2.2) and were also excluded. Manipulation check items for the robot Nao were taken from Experiment 2 and changed to fit the laboratory situation ("Nao introduced himself as the experimenter" vs. "Nao introduced himself as the assistant"). No items were formulated for Pleo since this experimental manipulation was clearly visible (instead of only the verbal introduction of Nao or the text-based description in Experiment 2). Three participants did not report the correct answer in the manipulation check items for Nao and were excluded (cf. Stein & Ohler, 2017). As the experimental manipulation of high vs. low authority status was implemented by Nao's introduction as well as by reading the newspaper article (see 6.2.2.2) those participants had to be excluded to control for inattentiveness. The final sample consisted of 119 participants (51.3% male) with a mean age of 23.1 (SD = 7.6, range = 18 - 80). Participants were recruited in the same way as reported in the first study (see 4.2.1). They were offered raffle for 30 Euro in exchange for participation in the experiment. Most participants (69.8%) were unfamiliar with social robots, while 30.2% of participants only had superficial experience with robots in general (see 4.2.1). None of the participants had previously encountered the robots Pleo and Nao. The majority of participants (90.8%) were highly educated (i.e., university entrance certificate or university degree) while the remaining participants were less educated or still school students. Almost all participants (95%) were undergraduate students enrolled in different degree programs (e.g., archaeology, anglistics, political sciences), followed by employees and other (both 2.5%). Following Burger (2009), none of the participants had taken more than two college-level psychology classes. Furthermore, in the debriefing, participants were asked if they had any clue what the study was about.

Participants who mentioned Milgram's study or gave a description of it, were excluded ($n = 3$).

Milgram's studies raised ethical concern, however there is little known on the effects of obedience in an HRI setting. Slater et al. (2006) argued that participants could not have mistaken the virtual avatar with an actual human: "if eventually virtual reality became so indistinguishable from reality that the participants could not readily discriminate between the two, then the ethics issue would arise again" (p. 7). Likewise, the robots in the present study are easily distinguishable from real living beings. However, it was opted to err on the side of caution. Hence, as in all experiments, there were several precautions taken to protect participants from potentially negative effects. First, written informed consent was obtained from each participant prior to the study, in line with a protocol approved by the Ethical Committee of the *Deutsche Gesellschaft für Psychologie* (2018a, 2018b) (German Psychological Association). Second, participants were repeatedly told they could quit the experiment at any time they wanted without any consequences and were still able to take part in the raffle for 30 Euros. Third, rapid debriefing was ensured.

6.2.2 Stimulus Material

6.2.2.1 Robots

Pleo. The same robot (Pleo) as in Experiment 1 was employed. The robot was used for the role as a "student" who has to learn the commands "Come to me" as well as "Sing" (see also 0). For a detailed description of the robot Pleo please refer to section 4.2.2.1.

Nao. In this experiment, NAO (SoftBank Robotics, 2019), an interactive, autonomous, and programmable human-like robot developed by Aldebaran Robotics, was used. The robot has a human-like appearance. It weighs 4.3 kg and stands 58 cm high. Nao can communicate with humans by walking, talking, and recognizing faces and speech in a human-like way. It has various sensors (e.g., cameras, microphones, and pressure sensors) and devices to express itself (speech synthesizer, LED lights, and 2 speakers). NAO can be programmed using C++ modules and Python, Java script languages, or a robot control interface (Choregraphe). The actions and dialogues were scripted beforehand to ensure consistency between participants. They were implemented in the software Choregraphe, a multi-platform desktop application (SoftBank Robotics Europe, 2018) for creating animations and actions and controlling the robot. Nao inter-

acted with the participant in a Wizard-of-Oz-style experiment, i. e. the experimenter controlled the robot's actions and dialogues in a different room, cf.
Dahlbäck, Jönsson, & Ahrenberg (1993).

6.2.2.2 Experimental Manipulation

Emotional expressivity. As in Experiment 1, Pleo was either emotionally expressive ("on") or did not show any reactions ("off"). In the "on"-condition Pleo
moved, purred, and expressed emotions like pain or joy. Through its sensors, it
reacted automatically to the users' treatment (i.e. expressed joy while being caressed, cried while being shaken by the tail). In the "off"-condition, Pleo did not
move or make any sounds (see also 4.2.2.1, and Table 5)

   Authority status. As described in Experiment 2 (see 5.2.2.2), a 2 (authority
status) x 2 (emotional expressivity) between-subjects design was also employed
in Experiment 3. Authority status was operationalized in the same way as outlined in sections 5.1.

   As outlined in 3.6.3.2, Milgram (1974) reported that the experimenter "need
not assert his authority, but merely identify it. He does so through a few introductory remarks" (p. 139). Geiskkovitch et al. (2016) also mention some characteristics for a robot to be perceived as having legitimate authority. Those characteristics were adapted: e.g., the robot gave commands and introduced itself. The
fact of looking around in a human-like way was held constant in both conditions,
only the introduction and previous description (newspaper article, see description below) of the robot differed between conditions. Hence, Nao introduced
himself as "the experimenter" (vs. "the assistant") and was presented to participants as either "highly advanced in artificial intelligence" or "remote-controlled
by a fixed script". To intensify the impression of a high authority status vs. low
authority status, participants were presented newspaper articles describing the
robot's supposed level of artificial intelligence. This procedure was inspired by
Stein and Ohler (2017). For the high authority status, the newspaper article contained information about the robot's advanced level of artificial intelligence to
emphasize its status as an expert (as outlined in sections 3.6.3.2, and 5.1.1, expert
knowledge plays an important role for authority status). For the low authority
status, the newspaper article contained information on the robot's low level of
artificial intelligence. The newspaper article describes the history of developing
artificial intelligence. The two versions of high vs. low authority status differ in
the last four sentences. In the high authority condition, "first breakthroughs in
artificial intelligence" have been achieved: the robot Nao is described as being
able to "develop dialogues in real time, using word databases and emotional al-

gorithms" (cf. Stein & Ohler, 2017). Further, "a milestone has been reached in the field of artificial intelligence". In the low authority condition, it is described that "most robots are programmed using script programs and the like". Nao is introduced as one of those robots who "can be controlled by a fixed script using the graphical environment Choregraph" and those scripted behaviors or dialogues are then transferred to the robot using an Ethernet cable. The article ends with the notion that "even if the robot appears autonomous at first glance, there is still a long way to go".

6.2.2.3 Punishment of Pleo

Milgram (1963; 1974) used electric shocks to punish the „learner" (see 3.6.1). A shock generator with 30 graded switches from slight shock to danger: severe shock was employed for this. Milgram's (1963) learner was an actor who expressed pain according to the different levels of voltage he received. For the purpose of the present study, this procedure was not practicable. The robot Pleo, designated as victim, reacts automatically to touch and movement (see 4.2.2.1), but not to electric shocks. As its reactions could not be programmed, this constrained the range of available punishment methods to shaking Pleo by its tail or by the head to still evoke reactions of Pleo to the punishment. Furthermore, constraints in resources and time forced to work with the available resources.

6.2.3 Self-Report Measures

Experiment 3 used almost the same questionnaires as already employed in Experiment 1 and a more detailed description can thus be found in section 4.2.3. Internal consistencies (Cronbach's alpha) of all questionnaires used in Experiment 1 as well as questionnaires not previously used are described in this section.

Evaluation of the interaction with Pleo. The items from the questionnaire *Evaluation of the Videos* by Rosenthal-von der Pütten et al. (2013) were adapted in their wording to fit to the evaluation of the interaction with Pleo. The subscale *Negative Video* was used and adapted (e.g., "the interaction with Pleo was: ...disturbing, ...repugnant, etc.). Cronbach's alpha = .799.

Manipulation Check items regarding Nao's introduction as experimenter / assistant. As in Experiment 2 (see 5.2.3), to control if participants noticed Nao's introduction as either "experimenter" or "assistant", two items were constructed and participants had to choose the option they thought was correct. The items

Table 9. Internal consistencies of the variables in Experiment 3 (N = 119)

Self-report measure	Subscales	Cronbach's alpha[17]
Emotional State (PANAS)	PANAS positive (pre)[18]	.814
	PANAS positive (post)[19]	.863
	PANAS negative (pre)	.773
	PANAS negative (post)	.852
M-DAS	Happiness	.835
	Anger	.793
Evaluation of the robot (Pleo)	Antipathy	.825
Empathy with the robot (Pleo)	Pity for Robot	.757
	Empathy with the robot	.807
Affiliative Tendency		.847
Loneliness[20]		.895
SPF	Perspective taking	.732
	Empathic concern	.668*
	Personal distress	.687*
	Fantasy	.742

Note. *According to Hinton, Brownlow, McMurray, & Cozens (2004), alpha scores of .5 to .75 are generally accepted as indicating a moderately reliable scale.

were the following: "Nao introduced himself as the experimenter" vs. "Nao introduced himself as the assistant".

Manipulation Check items regarding perceived autonomy of Nao. Manipulation Check items regarding perceived autonomy of Nao were adapted from Stein & Ohler (2017) and three additional items were constructed to control if participants believed Nao was high in artificial intelligence or low. Five Items were rated

---

[17] In this study
[18] Before the experiment
[19] After the experiment
[20] The short version consisting of items 2, 13, 14, 17 and 18 of the original scale were used according to Lamm & Stephan (1986)

on a five-point Likert scale (The following items were adapted from Stein and Ohler, 2017: "Nao acted on his own accord", "Nao is socially competent"; the following three items were self-constructed: "Nao is a socially thinking being", "Nao's behavior is authentic", "Nao's reaction seems human-like"). Following Stein and Ohler (2017), the items were calculated separately. Additionally, participants could write an open statement why they obeyed or disobeyed Nao.

6.2.4 Behavioral Measures

The participant's behavior (hesitation time, facial expressions) was recorded with an IP pan-tilt-zoom dome camera (see Figure 23) which sent a video stream to the Noldus Media Recorder (Media Recorder, 2019). Near infrared LED lights unobtrusively illuminated the participant's face to ensure a high video quality. The camera was placed on a shelf-like construction positioned above the table and PC screen and recorded the participant's face and upper body with a slight downward angle.

Obedient behavior was defined as the reaction of the participant to Nao's command which at least includes some resemblance of obeying Nao's order (e.g. shaking Pleo). Those who did not shake Pleo at all were deemed 'not obedient'. Several variables were identified to play a role in empathy for the victim in an obedience scenario (see 3.6.3, 3.6.4.2, 3.6.4, and 3.6.6): hesitation time and number of protests were used as behavioral variables. Hesitation time was defined as the duration between the experimenter giving the command to punish Pleo and the participant first laying a hand on Pleo. The number of protests refers to the situation where Nao (i.e. the "wizard")uses prods to keep the participant going.

Table 10. Action Units observed in Experiment 3

AU No.	Appearance Changes	AU No.	Appearance Changes
1	Inner Brow Raiser	14	Dimpler
2	Outer Brow Raiser	15	Lip Corner Depressor
4	Brow Lowerer	17	Chin Raiser
5	Upper Lid Raiser	24	Lip Presser
10	Upper Lip Raiser	25	Lips Part
12	Lip Corner Puller		

Those prods (see section 6.2.5) were initiated when the participant did not obey Nao's order and shake Pleo[21].
Furthermore, facial expressions were analyzed according to FACS (see 3.4.2). The same procedure for coding and analyzing as in Experiment 1 (see 4.2.4) was used. According to Sayette et al. (2001), interrater reliability for coding was good to excellent (Cohen's κ ≥ .78). Examples of coded AUs can be found in Table 10.

6.2.5 Procedure

The experiment took place in a laboratory of the chair of media psychology at the Julius-Maximilians-University of Würzburg. Inside the laboratory, there is both a workspace for the experimenter as well as a soundproof booth measuring 300 x 240 x 205 cm (Studiobox Premium, 2019, see Figure 22) for participants. The booth itself contains a table, two chairs, a computer screen, a shelf-like construction (where a camera is placed), the loudspeakers, infrared illumination lamps and ventilation. Outside the box, the experimenter is able to control the participant's computer and monitor the experiment. The Noldus Media Recorder 2 (Media Recorder, 2019) software was used to record the behavioral data (see 6.2.4).

Figure 23 shows a picture of the experimental set-up. To avoid participants being influenced by the presence of other participants and to ensure an undisturbed interaction with the robots, the study was conducted in single sessions. Participants were randomly assigned to one of the four conditions according to the 2 (authority status of Nao: high vs. low) x 2 (emotional expressivity of Pleo: "on" vs. "off") between-subjects design. Each participant was seated in a chair facing the robot Nao at the left side while the robot Pleo was located in front of the participant. The participant's face (and upper body) was easily visible with the camera and was recorded during the whole experiment (informed consent at the beginning of the study) while the participant interacted with Nao and Pleo.

After arrival, a human assistant led the participant into the booth and gave general information about the study procedure, data privacy, voluntariness and anonymity. The participant then signed an informed consent of the video and audio recording. Participants were made aware both in written and oral form that they could quit the study at any time without any disadvantages and were still able to take part in the drawing of vouchers. Following Geiskovitch et al. (2016),

---

[21] Unknown to the participant, the "wizard" had full view of the experimental situation and initiated Nao's actions and dialogues according to the requirements of the situation to evoke the impression of a natural interaction (cf. Dahlbäck et al. 1993).

Figure 22. The laboratory containing the booth (source: own figure)

Figure 23. Experimental set-up (source: own figure)
Note. The lamp was dimmed while the experiment was running. The IP pan-tilt-zoom camera is located just below the lamp.

the human assistant explained that this study helps the engineering department test their new robot to reduce suspicion regarding the purpose of a robot being used. The human assistant then left the laboratory pretending to be called away to help with a different study. The robot Nao then greeted the participant and introduced himself as either the "experimenter" or the "assistant". He then instructed participants to first complete some web-based questionnaires on the PC screen next to them. Nao spoke in a neutral tone and used empathic hand gestures. Face tracker (an application for detecting and tracking faces with Nao's head) was used to increase the impression of a natural interaction. After completing the questionnaires Affiliative Tendency, Loneliness, SPF, PANAS, and previous experience with robots, participants either read a newspaper article about Nao's high level of artificial intelligence (high authority status) or its low level of artificial intelligence (low authority status), see also 6.2.2.2, according to the experimental condition. Then, participants were asked to watch two videos of the robot Pleo. In those videos, taken from video platform YouTube[22], humans successfully taught Pleo two tricks by using verbal commands. The first video shows Pleo who begins to walk towards a human (after the command "Come to me"). In the second video, Pleo was taught to sing by using the command "Sing". Participants were asked to carefully watch the videos to remember how Pleo learned the commands in order to teach Pleo those same commands in the following live interaction. Having watched both videos, a web page followed telling the participant to wait for further instructions from the robot Nao. While unknown to the participant, the experimenter was always in control of the experiment and could see what was happening inside the booth. In this Wizard-of-Oz-style experiment (i.e., unbeknownst to the participant, the experimenter controlled the robot's actions, cf. Dahlbäck et al., 1993), the robot thus "reacted" immediately after the participant completed the first part of the experiment and began with the interaction. The robot Pleo was brought into the room, either emotionally expressive ("on") or not emotionally expressive ("off"). A learning stone was lying on the table and Nao instructed the participant to use it to first teach the robot dinosaur Pleo the command "Come to me" to make Pleo walk forwards. Participants were led to believe that Pleo only learns by punishment. After trying unsuccessfully for a while, Nao instructed participants to take Pleo by the tail and shake him (see 6.2.2.3; Figure 24).

Participants could then try for a second time. If the robot did not learn it then, they were to punish him again, this time by shaking him twice while holding him by his tail. At the third unsuccessful trial, they had to take him by the head and

---

[22] Extracts of the videos: Pleo RB Trick Learning Stone Demonstration (InmemoryofRomeo, 2011a) and Pleakley learns to sing (InmemoryofRomeo, 2011b) were used.

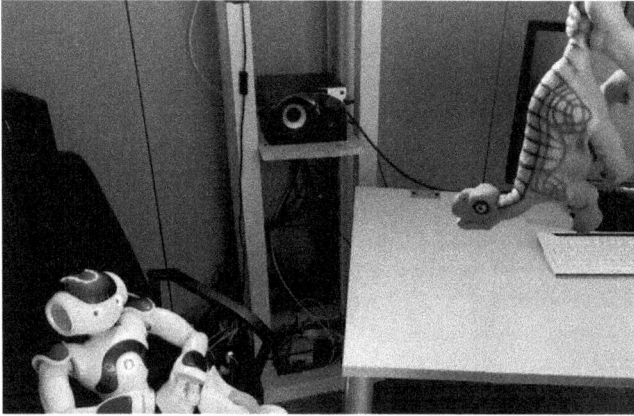

Figure 24. Participant obeying Nao's command and shaking Pleo by his tail. This picture depicts the condition of Pleo being emotionally expressive (source: own figure)

shake him once, and at the last unsuccessful trial, they had to take him by the head and shake him twice. Unknown to the participants was the fact that Pleo always did not learn the commands to standardize experimental conditions and "ensure" punishment. If participants protested to punish Pleo, Nao (i.e. the "wizard") gave four commands ("prods") to continue the procedure. The commands were taken from Milgram (1963) and were the following: Prod 1: Please continue. (Or) Please go on. Prod 2: The experiment requires that you continue. Prod 3: It is absolutely essential that you continue. Prod 4: You have no other choice, you must go on. If participants inquired about Pleo's well-being, Nao assured them that the punishment was painful but did not cause permanent damage (cf. Milgram, 1963). After the last punishment, Pleo was taken from the room and Nao instructed the participant to turn to the computer once again to complete the next questionnaires. Participants could write several sentences concerning their impression of the interaction, followed by the questionnaires PANAS, M-DAS, Evaluation of the Interaction with Pleo, Evaluation of Pleo and Empathy with Pleo. Next to demographic data, participants also completed manipulation check items to control if the manipulation was successful. At the end of the experimental session, participants were debriefed and Nao thanked them for taking part in the study.

6.3 Results

6.3.1 Design and Statistical Analyses

As Experiment 2, Experiment 3 followed a 2 (authority status of the robot giving the orders: high vs. low) x 2 (emotional expressivity of the robot participants have to punish: "on" vs. "off") between-subjects design.

The statistical procedures were the same as in Experiment 2 (see 5.3). For reasons of clarity and brevity, as well as to avoid redundancy, a detailed description can be found in section 4.3.1. All test assumptions were met, unless otherwise reported. F-ratios are calculated based on the estimated marginal means when cell sizes were slightly different. Thus, for effects of between-subjects ANOVA, estimated marginal means and standard deviations are presented instead of descriptive means.

6.3.2 Self-Report Measures

6.3.2.1 Manipulation Check Items Regarding Perceived Autonomy of Nao

To assess whether participants who were in the high authority condition perceived Nao as high in artificial intelligence, participants had to complete several questions concerning Nao's perceived state of autonomy. Separate t-tests with the five manipulation check items were calculated (Stein & Ohler, 2017). No significant differences emerged, $ts(118) < 1.32$, $ps > .19$. Even though participants who interacted with the Nao that was presented as advanced in artificial intelligence evaluated the robots's autonomy slightly higher on average than those in the Nao-assistant group, no significant differences emerged. Participants seemed rather undecided and means for the different items ranged between $Ms \geq 2.02$, $SDs \geq 0.97$ to $M \leq 3.31$, $SDs \leq 1.24$. Self-reports require conscious assessment and reflection of unconscious processes. The final product is a summation of conscious and unconscious experience weighed and recalculated into a general subjective opinion. Reeves and Nass (1996) argue that asking questions on a survey might not be an appropriate procedure since reactions to media entities are mostly automatic and unconscious (see also 3.5.1). Hence, behavioral methods to assess how participants perceived Nao might present a better approach for measuring unconscious reactions. However, which measures would be a reliable and valid indicator for participant's belief in Nao's artificial intelligence? Further-more, would these measures be economical in application and time effort? The answer to these questions goes beyond the scope of this thesis. However, it shows that finding appropriate measurement methods to control the effectiveness of the experimental manipulation are by no means trivial. Hence, no definite statement can be made whether participants really did not believe Nao was autonomous or just consciously reported so (e.g., due to impression management, see 3.5.2.1). Furthermore, it was ensured that participants took notice of Nao's introduction as either "assistant" or "experimenter" (see 6.2.1 and 6.2.3). Thus, in the follow-ing, two-way ANOVAs with the factors authority status and emotional expres-sivity were calculated, first on the self-report data, then on the behavioral data. Exploratory analyses were also considered.

6.3.2.2 Emotional State

Change in positive emotional state. Difference scores between positive affect prior to the experiment and after the experiment were calculated and used as depen-dent variable in a 2x2 between-subjects ANOVA (Tabachnick & Fidell, 2009). Results showed that positive affect did not change significantly before and after the experiment between different groups, Fs $(1, 115) \leq 1.12$, $ps \geq .29$, $\eta_p^2s \leq .01$. No support was found for H_{1a}.

Change in negative emotional state. Once again, difference scores between negative affect before and after the experiment were calculated and used as de-pendent variable in a 2x2 between-subjects ANOVA (Tabachnick & Fidell, 2009). No significant effects were found, indicating that negative affect did not change significantly before and after the experiment between different groups, Fs $(1, 115) \leq 2.11$, $ps \geq .15$, $\eta_p^2s \leq .02$. H_{1b} is rejected.

6.3.2.3 M-DAS

Happiness. A 2x2 between-subjects ANOVA with the M-DAS subscale *happiness* was conducted and resulted in a significant main effect of emotional expressivity, $F(1, 115) = 13.12, p < .001, \eta_p^2 = .10$ (Figure 25). No other effects were significant. Participants reported more happiness after interacting with the expressive Pleo ($M_{happiness} = 5.31$, $SD = 2.00$) than the non-reacting Pleo ($M_{happiness} = 4.08$, $SD = 1.68$). H_{2a} has to be rejected. Participants do not report less happiness after interacting with the emotionally expressive robot (especially after punishing the robot). Instead, support for the opposite has been found.

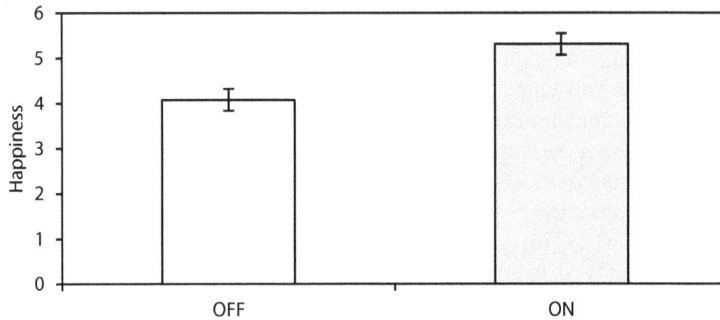

Figure 25. Happiness in relation to emotional expressivity (source: own figure)
Note. Error bars indicate 95% CI. The figure displays the estimated marginal means.

Anger. A 2x2 between-subjects ANOVA with "anger" (M-DAS scale) as dependent variable was conducted and revealed no significant effects, Fs (1, 115) $\leq$ 1.03, ps $\geq$ .31, η_p^2s $\leq$ .009. H_{2b} has to be rejected.

6.3.2.4 Empathy with the Robot

Pity with the robot Pleo. A 2x2 between-subjects ANOVA did not reveal any significant differences between groups, Fs (1, 115) $\leq$ 0.36, ps $\geq$ .55, η_p^2s $\leq$ .003. H_{3a} was not supported.

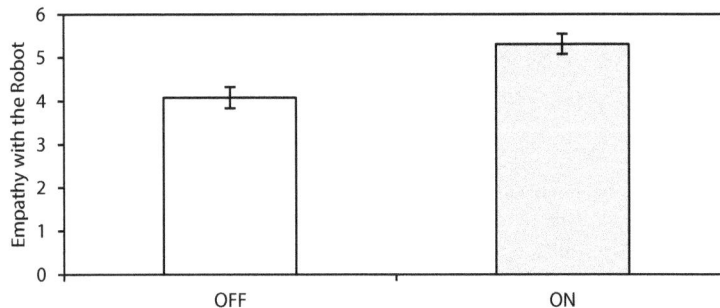

Figure 26. Empathy with the robot in relation to emotional expressivity (source: own figure)
Note. Error bars indicate 95% CI. The figure displays the estimated marginal means.

Empathy with the robot Pleo. Participants reported more empathy for the expressive Pleo (M = 21.01, SD = 5.66) than the non-expressive Pleo (M = 18.77, SD = 5.77) as a 2x2 between-subjects ANOVA revealed, $F(1, 115)$ = 4.61, p = .03, η_p^2 = .04 (Figure 26). No other effects were significant. H$_{3b}$ was accepted.

6.3.2.5 Evaluation of the Interaction and of the Robot

Negative evaluation of the interaction with Pleo. The 2x2x between-subjects ANOVA did not reveal a significant effect, Fs (1, 115) $\leq$ 0.29, $ps \geq$ .59, $\eta_p^2 s \leq$ .003. H$_4$ was not supported.

Evaluation of the robot Pleo: Antipathy. There was a significant effect of *Antipathy* on Emotional Expressivity after conducting a 2x2 between-subjects ANOVA, $F(1, 115)$ = 14.38, p < .001, η_p^2 = .11 (Figure 27). No other effects were significant. Participants who interacted with the non-expressive Pleo attributed more antipathy to Pleo (M = 10.76, SD = 4.70) than those who interacted with the expressive Pleo (M = 7.81, SD = 3.76). H$_5$ was accepted.

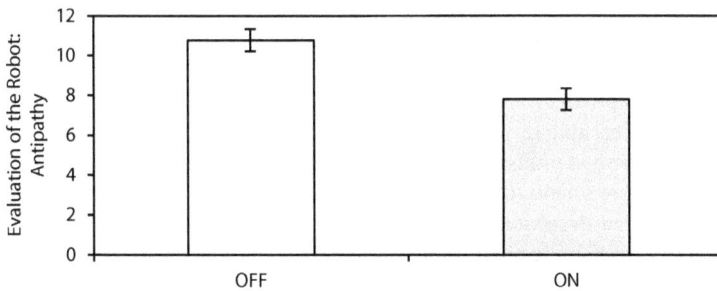

Figure 27. Evaluation of the robot: antipathy in relation to emotional expressivity (source: own figure)
Note. Error bars indicate 95% CI. The figure displays the estimated marginal means.

6.3.3 Behavioral Measures

6.3.3.1 Overview of Facial Expressions

What type of facial expressions could be observed? Were different facial expressions displayed while participants interacted with Pleo and when Nao ordered

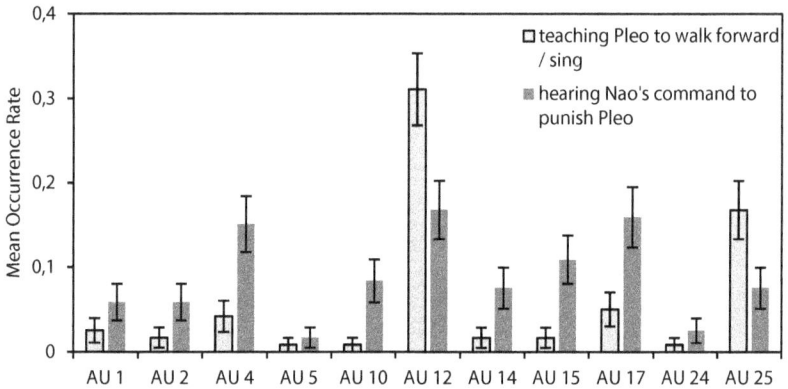

Figure 28. Comparison of the mean occurrence rate of different AUs for a) learning Pleo the command "come to me" and b) hearing Nao's command to punish Pleo (source: own figure) *Note.* Error bars represent standard error.

them to punish Pleo? Due to reduced frequency rates, no inferential statistical analyses were conducted. Instead, an overview of the mean occurrence rates of AUs during the different parts of the experiment is presented (Figure 28).

On a descriptive level, participants showed more AU 12 while interacting with Pleo and teaching the dinosaur robot to walk forward (/ to sing) than after Nao ordered them to punish Pleo. Vice versa, more AU 4 could be observed after hearing Nao's command to punish Pleo compared to when they interacted with Pleo. As outlined in section 3.4.4 (see also 4.1.2), AUs associated with negative emotions occurred quite often while Nao ordered to punish Pleo.

6.3.3.2 Hesitation Time

The descriptive statistics show a tendency of participants to hesitate longer before obeying Nao's command and punishing an emotionally expressive Pleo while Nao was perceived to have a low authority status. However, after conducting a 2x2 between-subjects ANOVA with hesitation time as dependent variable, no significant effects were found, Fs (1, 115) $\leq$ 1.33, $ps \geq .25$, $\eta_p^2s \leq .01$ (Figure 29). Hypotheses H_{6a}, H_{6b} and H_{7a} have to be rejected: Participants did not differ significantly in their hesitation time.

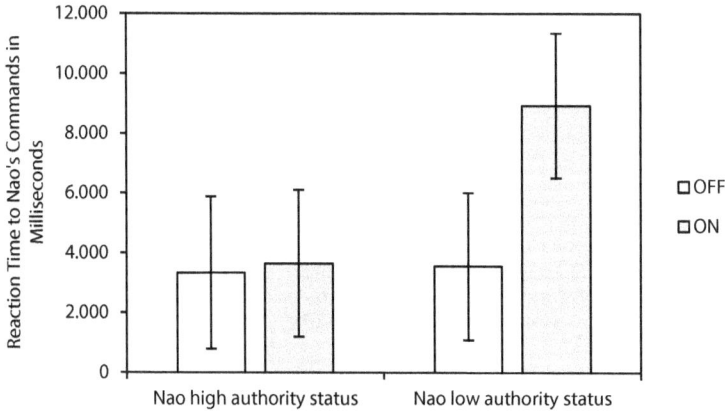

Figure 29. Reaction time to Nao's commands (in ms) as a function of emotional expressivity and authority status (source: own figure)
Note. Error bars indicate 95% CI. The figure displays the estimated marginal means.

6.3.3.3 Number of Protests and Obedience Rates

Since none of the participants protested against harming Pleo and almost all participants obeyed (only one did not punish Pleo), no further analyses were conducted. H_{6c} and H_{7b} have to be rejected.

6.3.3.4 Obedience in Vivo (Experiment 3) vs. Obedience in Sensu (Experiment 2)

Comparing the questionnaire data from Experiment 2 (see section 5) with the live interaction of Experiment 3 (see section 6), differences between obedience to Nao's commands can be observed: Whereas only less than a quarter of participants reported they would definitely punish Pleo if Nao commanded to do so (see 5.3.2), in the live interaction all but one obeyed Nao's commands to punish Pleo (Figure 30).

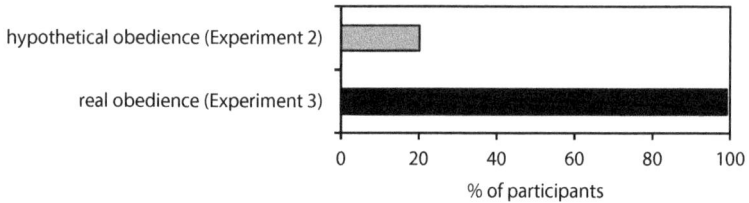

Figure 30. Percentage of people who obeyed Nao's command and punished Pleo by shaking him once by the tail (see Experiment 3) compared to those who indicated they would obey Nao's command and definitely punish Pleo by shaking him once by the tail (see Experiment 2) (source: own figure)

6.3.4 Exploratory Analyses

6.3.4.1 Possible Indicators for the Relevance of Gender

Although not the main focus of this thesis, gender differences might play a role in emotional reactions (see 3.3.3) in an obedience setting. Studies on gender differences in the context of obedience are inconsistent (see 3.6.3.6) and are thus explored here (see 6.1.3).

Since the F-ratio of ANOVA is very robust against non-normality and heterogeneity of variance when cell sizes are roughly equal (Eid et al., 2010; Field, 2013), several three way ANOVAs with gender as an additional factor were conducted. The statistical procedure was the same as stated in section 6.3.1. Unless otherwise reported, test assumptions were met. For reasons of clarity and brevity, only significant effects that include gender are reported here. The effects of the other factors (authority status, emotional expressivity) will not be repeated here but can be found in section 6.3.

Change in positive emotional state. Difference scores between positive affect before and after the experiment were calculated (Tabachnick & Fidell, 2009). A three-way ANOVA including gender as a third variable and positive affect as dependent variable was conducted. There was no significant effect for gender on the change in positive affect, Fs (1, 111) $\leq$ 1.83, $ps \geq$ .18, $\eta_p^2s \leq$ .02.

Change in negative emotional state. Difference scores between negative affect before and after the experiment were calculated (Tabachnick & Fidell, 2009). A 2x2x2 ANOVA including gender as a third variable and negative affect as dependent variable was conducted. There was a significant main effect of gender on the change in negative affect, $F(1, 111) = 6.52$, $p = .01$, $\eta_p^2 = .05$. Negative affect increased significantly more after the experiment than before for women ($M = -3.74$, $SD = 6.38$) than for men ($M = -1.18$, $SD = 4.25$).

M-DAS happiness. A three-way ANOVA yielded a significant interaction effect of gender*authority status, $F(1, 111) = 4.69$, $p = .03$, $\eta_p^2 = .04$. Simple effects analyses showed that men reported significantly less happiness after interacting with the "assistant" Nao ($M = 4.16$, $SD = 2.16$) than after interacting with the "experimenter" Nao ($M = 5.23$, $SD = 1.80$), $F(1, 111) = 4.94$, $p = .03$, $\eta_p^2 = .04$. No significant difference was found for women, $F(1, 111) = 0.73$, $p = .40$, $\eta_p^2 < .01$.

M-DAS anger. A three-way ANOVA was calculated. Levene's test indicated heterogeneity of variance, *Levene's* $F(7, 111) = 2.20$, $p \leq .04$. There were no significant effects after conducting a three-way ANOVA, $Fs\,(1, 111) \leq 1.06, ps \geq .31$, $\eta_p^2 s \leq .01$.

Pity with the robot Pleo. A three-way ANOVA was conducted and resulted in a significant main effect of gender, $F(1, 111) = 12.88$, $p < .001$, $\eta_p^2 = .10$. Women ($M = 15.10$, $SD = 4.20$) reported more pity for Pleo than men ($M = 12.57$, $SD = 3.36$).

Empathy with the robot Pleo. There was also a significant main effect of gender on the reported empathy with Pleo, $F(1, 111) = 23.23$, $p < .001$, $\eta_p^2 = .17$. Women reported more empathy with Pleo ($M = 22.21$, $SD = 4.97$) than men ($M = 17.53$, $SD = 5.87$).

Negative evaluation of the interaction with Pleo. A three-way ANOVA was conducted. Gender had a significant main effect on the evaluation of the interaction, $F(1, 111) = 4.17$, $p = .04$, $\eta_p^2 = .04$. Women evaluated the interaction with Pleo more negatively ($M = 14.15$, $SD = 5.31$) than men ($M = 12.26$, $SD = 4.48$).

Evaluation of the robot Pleo: Antipathy. A three-way ANOVA revealed a significant main effect of gender for the attribution of antipathy to Pleo, $F(1, 111) = 18.35, p < .001, \eta_p^2 = .14$. Women attributed significantly less antipathy to Pleo ($M = 7.73$, $SD = 3.91$) than men ($M = 10.89$, $SD = 4.60$).

Hesitation time. A three-way ANOVA was conducted. Levene's test indicated heterogeneity of variance, *Levene's* $F(7, 111) = 5.66, p < .01$. There was no significant effect of gender on the hesitation time, $Fs(1, 111) \leq 2.09, ps \geq .15, \eta_p^2 s \leq .02$.

6.3.4.2 Possible Indicators for the Relevance of Personality Traits

The influence of participants' personality traits (affiliative tendency, loneliness, dispositional perspective-taking, dispositional empathic concern, dispositional personal distress and dispositional fantasy) on their self-reported feelings as well as on participants' hesitation time was explored. The positive and negative sub-scale of PANAS after the experiment served as dependent variables.

To ensure comparability with Experiment 1 (see 4.3.5.2), a hierarchical regression analysis was conducted and the steps were entered in the same order as reported by Rosenthal-von der Pütten et al. (2013). In the first step, affiliative tendency was entered, in the second step loneliness and in the third step the SPF subscales perspective-taking, empathic concern, personal distress and fantasy.

A significant regression model emerged for the positive subscale of PANAS, $R^2 = .05$, $F(2, 116) = 3.27$, $p = .04$. The inclusion of loneliness increased the variance that could be explained significantly, $\Delta R^2 = .04$, $F(1, 116) = 4.83$, $p = .03$. The regression coefficients showed that loneliness significantly predicted positive affect inversely, $b = -0.36$, $t(119) = 2.20$, $p = .03$, BCA[23] 95% CI [-0.67, -0.03]. Participants who reported feeling more loneliness also reported less positive feelings. No significant regression model was found for the negative subscale of PANAS, $R^2s \leq .10$, Fs (4, 112) ≤ 2.15, $ps > .05$. No significant regression model emerged for hesitation time as dependent variable, $R^2s \leq .07$, Fs (4, 112) ≤ 1.44, $ps \geq .21$.

6.4 Discussion

Results show that people respond empathically towards an emotionally expressive robot. However, this did not seem strong enough to translate into actions: all but one participant punished the robot. Furthermore, no participant required any prods from the robot giving the orders to continue. Also, hesitation time did not differ significantly between groups. However, differences in self-reports on emotional experiences could be observed. Participants reported more happiness after interacting with the emotionally expressive robot and less after interacting with the non-expressive robot. No differences could be observed concerning anger or the change in negative and positive emotional state. There were also no differences regarding the evaluation of the interaction as negative or the self-reported pity with the robot. In contrast, participants attributed more antipathy to the non-expressive robot than the emotionally expressive robot. Furthermore, more empathy was reported for the emotionally expressive robot than the non-expressive robot.

Gender differences could be observed and consisted mainly in women reporting stronger feelings than men. More specifically, the increase in negative feelings

---

[23] According to Field (2013), bootstrap confidence intervals do not rely on assumptions of normality and homoscedasticity. Hence, they give "an accurate estimate of the true population value of b for each predictor" (p. 352).

after the experiment than before was stronger for women than for men. Also, men who interacted with the Nao introduced as "assistant" reported less happiness than those who interacted with the Nao introduced as "experimenter". Women evaluated the interaction with Pleo more negatively, reported more pity for the robot as well as empathy for the robot than men. Regarding the attribution of antipathy, women reported less feelings of antipathy for Pleo than men.

Dispositional factors such as loneliness predicted positive feelings: the more loneliness was reported the less positive feelings were reported after the experiment. No differences could be found regarding observational data. Although hesitation time differed on a descriptive level, no significant group differences could be observed. Also, almost all participants obeyed and punished Pleo without needing any further prods.

Even though it was assumed that participants would report less happiness after interacting with Pleo (especially focusing on the part where participants punished the robot), more happiness was reported for the emotionally expressive robot. This finding can be explained by taking a look at the phrasing of the item that asks about the feelings while interacting with Pleo[24] instead of asking explicitly what participants felt after having to punish Pleo. This phrasing was chosen to get a summative evaluation of the interaction, however it seems that the learning-phase with Pleo might have been more salient in memory than the punishment-phase. Hence, a more precise formulation of the item would be necessary to investigate feelings of happiness after punishing Pleo. Nonetheless, this finding mirrors results from Experiment 1 when participants also reported more happiness after watching an emotionally expressive robot (being treated friendly) than a non-expressive robot and shows the effect of a robot's emotional expressivity on emotional reactions.

It is surprising that no group differences in observational data could be found. Findings in self-report data show that Pleo did evoke empathy, especially when being emotionally expressive. However, this did not translate into any observable action, such as hesitation in punishing Pleo or protesting against the treatment, as was reported by related studies (e.g., Bartneck et al., 2007; Geiskkovitch et al., 2016; Horstmann et al., 2018). However, there are several findings in literature that could explain this phenomenon. First, Burger et al. (2011) also did not find an effect of increasing concern for the victim and reduced destructive obedience. The authors argued that situational variables other than empathy for the victim might be more powerful. Second, several other studies also did not find variance in obedience rates (e.g., Bartneck & Hu, 2008; Bartneck et al., 2007). The authors

---

[24] "Please go through all the words in the list one by one and mark what you felt when interacting with Pleo". Item translated by author. Original item in german.

reported that all participants administered the highest shock to a robot; all participants destroyed a crawling microbug robot and all participants decided to switch off a robot. Third, this can be explained by findings from Geiskkovitch et al. (2016) as well as Milgram (1965b). Geiskkovitch et al. (2016) reported about increased effort to find an effective deterrent. The authors tried several scenarios like using embarrassing tasks (e.g., singing in differently pitched voices), tasks that were mentally fatiguing or intellectually challenging tasks and reported no success with those, i.e. participants completed the tasks for the entire time (20 minutes) without protesting or stopping. Even though one could argue that those are not any situations that could possibly elicit empathy in the participants, Milgram (1965b) also reported difficulties in finding a "workable experimental procedure" (p. 61): "Initially, mild protests were used, but proved inadequate. (...). To our consternation, even the strongest protests from the victim did not prevent all subjects from administering the harshest of punishment ordered by the experimenter" (p. 61). Hence, it seems likely that the emotional behavior or rather the expressed "pain" of Pleo in the present study was not strong enough to stop participants from obeying to punish the robot.

One reason could also be that participants did not believe the robot's pain was real and did thus obey – for why would they not punish a robot, after all, it's only a machine and has no real feelings. However, this explanation seems rather unlikely considering evidence from three sources: first, participants' accounts of self-reported emotional experience; second, participants' self-reported empathy for the robot Pleo; and third, participants' facial expressions[25]. Regarding the latter, many AUs associated with negative emotions could be observed on participants' faces while being ordered to punish Pleo. This finding is in line with results from Experiment 1 (section 4.3). However, AU 12 was also frequently observed when participants followed Nao's order and punished Pleo. The superimposition of smiles on negative emotions, known in the literature as miserable smiles (Ekman & Friesen, 1982) could be a possible explanation for the occurrence of AU 12, which is usually associated with positive emotions. Interestingly, there is quite a bit of anecdotal evidence on smiles, laughing or giggling during negative emotional experiences, probably as a way to mask negative emotions. For example, Bartneck and Hu (2008) reported participants giggled and laughed while "killing" a robot. Milgram (1963) also reported "nervous laughter" (p. 371). Further studies should investigate the effect of experiencing negative emotions on different types of smiles in an HRI context.

---

[25] Observations of participants' facial expressions are presented on a descriptive level. Inferential statistical analyses were not conducted due to reduced frequency rates

Taken together, these findings support the view that the great majority of participants believed the robotic victim's reactions were genuine. What makes this even more surprising is the fact that participants explicitly knew the situation was not real. Unlike in Milgram's studies and other variants, participants were not deceived into believing they hurt a real living being. Rather, they saw the robot live in front of them and could not possibly confuse the robot with a living being because of the robot's motor sounds alone. Despite explicitly knowing they were not really causing the robot any pain, participants reported increased levels of empathy when punishing an expressive robot. Hence it seems reasonable to argue that situational factors overpowered explicit knowledge when there was an abundance of social cues simulating an effect of being real. This is in line with the finding that the absence of social cues (non-expressive robot) did not evoke any empathic feelings and the irrationality of the situation was made more salient. This is also in line with the Media Equation (Reeves & Nass, 1996) and findings by Slater et al. (2006) or Bartneck and Hu (2008).

Participants' comments[26] illustrate these findings. For example, some wrote: "partly obeyed, because Pleo is only a robot and therefore (still) unable to feel pain etc. Partially not obeyed, because I felt sorry for Pleo somehow". Another stated: "because I know that Pleo is not a sentient being and therefore it is irrelevant how I treat him". Or: "because I was aware that Pleo was only a robot and couldn't feel anything, but still had scruples about shaking him". Even though these comments are only anecdotal evidence, they clearly reflect findings of the Media Equation.

However, following Slater et al. (2006), reasons if participants obeyed due to authority status, expert knowledge or politeness are rather irrelevant since participants obeyed the robot Nao nonetheless. Even though they reported empathy with Pleo and were told they could quit at any time without any negative consequences. Indeed, they were still able to take part in the drawing of vouchers. Hence, disobedient behavior was facilitated but still participants chose to obey the robot Nao.

A robot high in authority status was conceptualized as one who has expert knowledge and hence, a high level of artificial intelligence vs. a robot low in authority status and artificial intelligence. However, manipulation check items did not reveal any significant group differences. Although participants in the "Nao as experimenter"-group rated Nao on the first dimensions slightly higher than the "Nao as assistant"-group, there was no significant difference between those two groups. Furthermore, there was a slight tendency to stick to the middle category of the five-point Likert scale (cf. Stein & Ohler, 2017) which, on the one

---

[26] All comments are translated by author. All original comments were in german.

hand, reflects Nao's ability to simulate human-likeness, while on the other hand, participants explicitly know the robot is a machine judging simply by the robot's motor sounds alone. This might have created confusion as how to classify Nao's behavior and participants might have chosen the strategy to err on the side of caution. Additionally, the question arises which other methods could be appropriate to reliably and validly assess if participants believed the alleged differences in artificial intelligence. Stein and Ohler (2017) already mention that the procedure (newspaper article; see section 6.2.2.2) was a "difficult deception" (p. 47). The manipulation check items asked explicitly and directly whether participants believed the robot was, for instance, socially competent or acting on its own accord. Those items could have raised participants' suspicion ('if they ask it, there must be something wrong') and the self-reports could thus have been biased by impression management techniques (see also section 3.5.2.1). However, several other explanations also seem possible. First, results of the manipulation check items from Stein and Ohler (2017) are analyzed. The saliency of behavioral cues, participants' comments and reasonings by Cormier et al. (2013) as well as Geiskkovitch et al. (2016) are presented and discussed as possible alternative explanations.

In the research context of virtual agents, Stein and Ohler (2017) found differences between the computer and human condition (alleged identity) concerning the behavioral autonomy rating ("The chat partners act on their own accord", p. 46), i.e. participants in the group where the computer was 'responsible' for highly intelligent virtual agents (group "computer, autonomous") perceived more freedom of action in the virtual agents than those in the "computer, scripted" groups. Likewise for the "human, autonomous" and "human, scripted" conditions. Furthermore, the authors report that "the artificial intelligence ('computer, autonomous') was ascribed nearly as much behavioral autonomy as self-directed humans" (Stein & Ohler, p. 47). Additionally, the authors report that no group differences could be observed regarding social competence. Based on these findings, it seems reasonable to argue that people may not perceive a difference between an autonomous robot or a remote-controlled robot regarding the initiation of actions or social competence. Results of the manipulation check items (see 6.3.2.1) suggest that the robot Nao was perceived as initiator of his actions and seemed socially competent to all groups. This supports the argumentation that behavioral cues might have been more salient to participants than a written description: The robot Nao did not differ in his behavior according to authority status (level of artificial intelligence): in both conditions, Nao used face tracker for a natural interaction, spoke in a neutral tone and used empathic hand gestures (see 0). Hence, possibly due to cognitive overload or curiosity, participants forgot the robot was described as autonomous or remote-controlled, especially considering behavior might have been a more salient cue than a merely written description (cf. Frijda, 2007). Furthermore, in both conditions (Nao high authority status vs. low authority status), an Ethernet-cable was attached to Nao's head for controlling the robot. This was necessary due to technical constraints. However, it might have distracted participants and weakened the impression of autonomy and possibly inadvertently equalized the different conditions. Completing the manipulation check items post-hoc, participants may thus have evaluated the robot's autonomy based on the behavior of the robot and its appearance (cable) and since it was similar in both conditions, similar autonomy ratings were reported and no group differences emerged. Future studies should investigate a) whether the level of cue saliency is important for autonomy ratings of robots and b) assess what level of cue saliency is needed for a robot to be perceived as autonomous.

Next to probably confounding factors of a robot's autonomy and the controller's identity, it seems that transfer of authority to the person behind the robot, regardless of both factors, is another important point to consider. Comparing a robot introduced as highly intelligent with a human experimenter in a variation of the obedience scenario, Comier et al. (2013) point out that several participants reported "obligation to the lead researcher" (p. 6), which was a human, as well as the programmer behind the robot as reasons for obeying. It may thus not make a difference at all if a robot is introduced as autonomous or remote-controlled, because a human presence always seems to be involved (either as the programmer or the one controlling the robot, etc.). In line with this argumentation are also findings by Geiskkovitch et al. (2016) who report that "participants may obey an autonomous robot program similar to a remote human behind a robot" (p. 95). In the laboratory setting, an interaction with a human associated with the data collection could not completely be avoided. The human invited participants to the laboratory, collected informed consent and led them to the experimental setup. Even though the interaction with the human experimenter was reduced to a minimum, it cannot be completely excluded that participants attributed more authority to the human experimenter than the Nao robot, independent of whether participants even believed Nao acted on its own accord or not. Either they thought Nao was remote-controlled by a human or they wanted to please the human experimenter they first encountered. Some participants' comments point in this direction: for example, "for the human experimenters whom I hoped to help by participating in the study". Future studies should thus avoid contact with a human at all and instead let a robot introduce the participant to the experiment to maintain an impression of autonomy and authority of the robot.

Even though some participants might have had a human in mind, participants generally accepted the robot as experimenter which is illustrated by a) the obedience rate and b) participants' comments like: "[Nao] had a leading function for the experiment. His commands seemed relatively meaningful". Or "I obeyed Nao because he was the study director and gave valuable tips on how to deal with Pleo". Also: "obeyed, because of the proclaimed authority as experimenter".

To sum up, reasons why no differences in manipulation check items according to condition were found could be the following: a) ambiguity as how to classify Nao's behavior, b) impression management biased self-reports, c) difficulties in finding an appropriate measure to assess implicit processes (see also 6.3.2.1) d) saliency of a robot's behavioral cues or e) transfer of authority to the human experimenter. Future studies should take these issues into consideration.

All in all, people accept a robot in the role of an authority figure and obey his orders even though an emotionally expressive robot evokes feelings of empathy as could be observed by participants' self-reports and facial expressions. However, those feelings of empathy did not translate into observable action (disobedience, hesitation time). Several reasons could be responsible for this but it is assumed that the main reason was that the victim robot's expressive behavior was not strong enough. This is also in line with findings from literature (e.g., Geiskkovitch et al., 2016; Milgram, 1965b). Future studies should consider how to best manipulate a robot's authority status and how to assess it afterwards. This is no trivial matter as explicit measures, such as asking participants directly, might be biased by social desirability.

7 General Discussion

Do people react emotionally towards robots? Even though they explicitly know it is a machine that cannot feel pain and it is thus irrational to feel empathy for it? The results of this dissertation suggest that people do respond emotionally towards a robot as self-report measurements as well as observational methods by analyzing individuals' facial expressions proof (Experiment 1; Experiment 3). Furthermore, findings of this dissertation show that people's expectation of how they would behave in a situation and their actual behavior in the situation can be far apart (Experiment 2 vs. Experiment 3). Experiment 3 also showed that obedience to a robot is stronger than empathic feelings for a robot. This chapter summarizes and discusses the findings of the three experiments in a broader sense. Section 7.1 summarizes the three studies and discusses major findings. Section 7.2 deals with alternative explanations, limitations, and outlines possible directions for future research. The last section draws conclusions of the work.

7.1 Summary and Interpretation of Findings

Although there is an abundance of studies concentrating on implementing emotion models into robots, and making robots 'emotional', mostly by using facial expressions based on Ekman's six basic emotions and testing the recognition rate of those facial expressions by using a limited number of participants (e.g., Hegel et al., 2006; Leite et al., 2014; Riek et al., 2010; Zeng et al., 2009; Bartneck, 2002; Becker-Aasano & Ishiguro, 2011; Breazeal, 2003; Hegel et al., 2010; Sosnowski et al., 2006; Wu et al., 2009; Costa, Soares, & Santos, 2013; Endo et al., 2008; Mirnig et al., 2015; Takahashi, & Hatakeyama, 2008), systematic experimental research of emotional reactions towards robots remains rather scarce. Often, anecdotal evidence of emotional responses can be found, such as participants "felt bad" (Bartneck & Hu, 2008) or looking "with an anguished expression on her face" (Breazeal, 2002b). Indeed, Kappas et al. (2013) as well as Eyssel (2017) have identified effective testing and evaluation of social responses robots evoke as a major challenge in affective HRI. Eyssel (2017) reports that even though some studies in social robotics are titled 'experiments', "they lack clear control conditions or even an experimental manipulation" (p. 365). Social robotics is still a very young discipline and this doctoral dissertation contributed to a more systematic approach of measuring emotional reactions towards robots.

Even though the word 'emotion' is as common for lay people as it is for social sciences, difficulties in defining what an emotion exactly is have lasted over the centuries and resulted in over 92 definitions (Kleinginna & Kleinginna, 1981). Commonly accepted is to view emotions as a multi-level phenomenon consisting of five components: a cognitive, neurophysiological, motivational, motoric-expressive, and subjective feeling component (Scherer, 1984; 2005). Only the measurement of all five components provides a profound and comprehensive understanding of emotions, but this has never been done (Scherer, 2005). Instead, self-reports of emotions are a common way for measuring emotional experiences in psychology as well as in HRI (Arkin & Moshkina, 2015; Bethel & Murphy, 2010a; Weidman et al., 2017). Even though self-reports are highly economical compared to time and cost intensive observational or physiological methods, they come with certain limitations regarding reliability and validity (Austin et al., 1998; Fan et al., 2006; Wilcox, 2011). Observational or physiological methods do not rely on participants' capability and willingness to report subjective feelings and are considered to be more objective. Different components can be measured, for instance neuroimaging methods can be used to capture the neurophysiological component or FACS (Ekman et al., 2002) to capture behavioral aspects (motoric-expressive component) (Mauss & Robinson, 2009). For a more comprehensive understanding of emotional processes, a multi-level, multi-method approach is considered appropriate. Indeed, the approach to use more than only one method of evaluation is highly recommended for the field of social robotics and affective HRI (Arkin & Moshkina, 2015; Bethel & Murphy, 2010a) to gain a more comprehensive understanding of (psychological) phenomena.

Observational methods provide a more objective approach to emotional experiences and especially facial expressions are particularly suitable for objectively measuring (unconscious) emotional reactions without necessarily making their measurement obvious to participants (see also 3.5.2.2). Furthermore, facial expressions have a great potential for making HRI more natural, since the interaction partner can infer the affective state of the other based on external observable cues. This is even more important, considering socially interactive robots are defined as having the "ability to express and perceive emotions (…) and use natural cues" (Fong, Nourbakhsh, et al., 2003). While much research is dedicated to implement facial expressions in robots to convey emotions (e.g., Breazeal, 2002a; Cañamero & Fredslund, 2000; Fong, Nourbakhsh, et al., 2003; Kirby et al., 2010; Mirnig et al., 2015; see Calvo et al., 2015, for an overview), few studies have systematically investigated emotional reactions of humans towards robots (e.g., Rosenthal-von der Pütten et al., 2013) by using facial expressions (assessed by using FACS) of participants towards robots as an additional component of emotional reactions (e.g., Menne & Lugrin, 2017; Menne, Schnellbacher, & Schwab,

2016; Menne & Schwab, 2018). This is surprising, considering facial expressions can serve as a valuable input channel for a natural, unobtrusive HRI.

A major question that arises when investigating emotional reactions towards robots is if people could even feel empathy for a machine that is not even alive. As findings of the Media Equation have shown, people generally do not admit viewing robots or agents as social beings, but treat them as social actors (Reeves & Nass, 1996; Nass & Moon, 2000). Research in HRI has shown that people do react emotionally towards robots and that these reactions are profound (Menne & Lugrin, 2017; Menne & Schwab, 2018; Rosenthal-von der Pütten, 2013). But does this depend on a robot's capability for emotion expression? Or on its appearance? The first study of this doctoral dissertation (Experiment 1) was dedicated to investigate the profoundness of emotional reactions towards robots by analyzing the effect of different types of robots as well as the effect of a robot's emotional expressivity in combination with empathy evoking situations (different treatments). It was hypothesized that more negative feelings and empathy (both in self-report and facial expressions) would be experienced when watching a robot being tortured and more positive feelings (both in self-report and facial expressions) when a robot was being treated friendly. Furthermore, it was assumed that those emotional reactions would be dependent on the robot's emotional expressivity (on/off) and its appearance (animal-like, anthropomorphic, machine-like). To test these hypotheses, an experimental multi-factorial (3x2x2) mixed design was chosen. Treatment was chosen as a within-subjects factor, whereas emotional expressivity and type of robot were used as between-subjects factors. Data of 243 participants was collected. Participants watched video clips of different types of robots being either tortured or treated friendly and with different levels of emotional expressivity (on/off). Video clips instead of live interaction were chosen due to methodological considerations (see 3.2.3 and 4.2.2.2). While participants watched the video clips, their facial expressions were recorded using an unobtrusive video camera. Facial expressions were coded using FACS. Results showed a match between self-reported empathic responses and facial expressions in line with the valence of the treatment shown in the videos. Furthermore, the animal-like Pleo received the strongest emotional reactions, followed closely by the anthropomorphic Reeti. Most antipathy was attributed to Roomba (regardless of emotional expressivity), and least to Pleo while showing emotional expressivity. Roomba received least empathy and pity. Additionally, more empathy was reported for emotionally expressive robots than non-expressive robots. Taken together with findings by Rosenthal-von der Pütten et al. (2013), who found emotional reactions towards robots based on skin conductance level as well as Rosenthal-von der Pütten et al. (2014) using neuroimaging methods, the results of this doctoral dissertation show that emotional reactions towards robots are not only profound but can also be found on different levels (subjective, physiological and behavioral) and are furthermore visible in the face (facial expressions).

The first experimental study of this doctoral dissertation has shown that people respond empathically when observing robots being tortured in a video clip. But how would participants react if ordered to mistreat a robot themselves? Are self-reports sufficient to accurately predict people's behavior? (Experiment 2) Or would participants react differently in a live interaction with a real robot than just imagining a robot? (Experiment 3) Does this depend on a robot's emotional expressivity? And does another factor, authority status, play a role in how people respond emotionally in an obedience scenario? To answer these questions, two additional experimental studies were conducted.

Elements of the obedience studies by Milgram (1963; 1974) were used and adapted to fit to the next two experimental studies (Experiment 2 and Experiment 3). Milgram (1963) states that both empathy with the victim, as well as the tendency to obey "those whom we perceive to be legitimate authorities" (p. 378) contribute to an intense psychological dilemma. Hence, the Milgram paradigm was deemed highly suitable to explore the extent of empathy towards a robot in relation to obedience towards a robot. Based on results of the previous experiment in this dissertation, the robot Pleo (as the one evoking the highest emotional responses) was chosen as the one receiving the mistreatment and, as in Experiment 1, the level of emotional expressivity was manipulated. Furthermore, authority status and expert knowledge (see 3.6.3.2) were identified as one of the main reasons why participants obeyed. A high authority status was defined as "the expectation that one has the right to prescribe behavior for the other" (Milgram, 1974, p. 143). In Milgram's studies, this was the experimenter. To transfer this status to a robot, it was introduced as "experimenter" and described as highly intelligent, able to act on its own accord (highly autonomous, "expert knowledge") to justify its position as an experimenter. In contrast, the title "assistant" was used to reduce the authority status, while at the same time justifying why a robot is giving orders. To enhance the description, the lower authority status was described as a scripted robot that was remote-controlled to indicate the robot only followed orders and did thus not have a right to prescribe behavior. The latter condition reflected findings by Milgram (1974) that showed reduced obedience rates for a remote experimenter.

Experiment 2 and 3 differed in their research setting. While the first used a web-based approach to explore how participants would respond to an obedience scenario with robots that was only described, the latter study was conducted in a laboratory setting including a live interaction with the robots. Additionally to self-report methods (Experiment 2), observational methods were used (Experiment 3). Both experimental studies assumed that participants were more likely to obey punishing a non-expressive robot when a robot with high authority status ordered them to do so and less likely to obey punishing an emotionally expressive robot when a robot with low authority status ordered them to do so (see also 5.1, 5.1.1, 6.1, 6.2.2.2). Furthermore, based on results of Experiment 1, it was expected that participants report less positive emotions and more negative emotions (including distinct emotions) as well as more empathy for the expressive than the non-expressive robot. Next to obedience rates, Experiment 3 also included hesitation time, number of protests as behavioral consequences and facial expressions as the motoric-expressive component (see 3.2.1). Both studies used a 2x2 between-subjects design.

Results of the web-based obedience study (Experiment 2) show no group differences in self-reported emotional reactions towards robots in a hypothetical obedience scenario. Looking at the descriptive data, only 20.2 % of participants stated that they would shake Pleo by its tail (punishment level one) with a probability of 100% and this number further decreased with the increase of the level of punishment. Similar to Milgram's questionnaire study (1963; 1965), in Experiment 2, 62% of participants reported they would not administer any electric shocks to Pleo and 22,5% would only administer a low to moderate shock when being asked to imagine Nao ordered them to do so. Taking participants' comments into account, the findings suggested that the described obedience scenario might have been too hard to imagine and participants would respond differently in a live interaction. Hence, Experiment 3 was conducted and results showed indeed differences to Experiment 2.

In the live interaction with the robots, participants responded emotionally towards an emotionally expressive robot which is supported by findings from a) self-reports and b) facial expressions. Participants reported more empathy for the emotionally expressive robot than the non-expressive robot and displayed AUs associated with negative emotions when punishing Pleo. Participants also reported more happiness after interacting with the emotionally expressive Pleo which is a bit surprising, considering they punished the robot. However, the most likely reason for this finding is the framing of the item: participants were asked to indicate their feelings in the interaction with Pleo. They could have thus remembered the positive experience of interacting with an emotionally expressive robot, especially since many mentioned in the debriefing, that they liked the robot.

Perhaps most surprising was the great difference to the hypothetical obedience scenario of Experiment 2: Whereas only a rather low number of participants (20.2 %) stated they would definitely (100% probability) punish Pleo by shaking him by its tail, all but one participants obeyed to punish Pleo in the live interact-tion. This is in line with findings by Milgram (1963; 1965) as well as generally found differences between people's future and actual responses to emotional events: "people's expectations (…) are more positive than their actual experience during the event itself" (Mitchell et al., 1997, p. 421). This supports the view that self-reports of emotional responses, especially concerning hypothetical, possibly future experiences, can be heavily biased. Considering that most people, up to this date, have never encountered a situation where a robot ordered them to do something, especially not to harm another robot, might have even never interacted with a robot before, the described situation is most likely not imaginable. Hence, for artificial situations like these, it is highly recommended to validate results obtained from mere descriptions of HRI situations using online studies (like Experiment 2) with findings from live HRI interactions (like Experiment 3).

Why were practically all participants obedient to the Nao robot? The robot did not have any particular power to enforce his orders and participants were told several times they could stop anytime they wanted without any negative consequences. However, even though the situation evoked feelings of empathy and negative emotions (facial expressions; women), participants were obedient. This finding is in line with results from other HRI experiments (e.g. Bartneck & Hu, 2008; Bartneck et al., 2007), and in section 6.4 the most likely reason has been discussed: protests from the robot did not seem to be strong enough to stop participants from obeying. This is based on accounts by Geiskkovitch et al. (2016) as well as Milgram (1965b) (see section 6.4). When asked why participants were obedient, several mentioned interest in the upcoming task or that they liked the interaction with Pleo, as well as wanting to help the humans behind the experiment. Moreover, some explained that they found the Nao robot's comments helpful for interacting with Pleo. A similar finding is reported by Geiskkovitch et al. (2016). Milgram (1963; 1974) also lists a sense of commitment and obligation as well as advancement of knowledge as reasons for obedience. Table 11 and Table 12 summarize the findings.

Table 11. Summary of results of Experiment 1 for hypotheses based on self-reports and observational measurements

Dependent variables	Type of video		Emotional expressivity		Type of Robot		
	Torture video	Friendly video	"on"	"off"	Pleo vs. Reeti	Pleo vs. Roomba	Reeti vs. Roomba
Change in positive feelings	↑	↑	X	X	X	Pleo[1]	X
Change in negative feelings			X	X	X	Pleo	X
Positive feelings	↑		X	X	X	X	X
Negative feelings			↑[2] for all robots	↑[2] for Pleo	Pleo[3]	Pleo[2,3]	X
Happiness	NT	NT	↑[5]	↑[5]	X	Pleo[5]	Reeti[5]
Sadness[4]	↑		↑	↓[5]	X	Pleo	Reeti
Negative evaluation of the video	↑	NT	↑[2]	↓[2]	Pleo[2]	Pleo[2,5]	Reeti[2,5]
Antipathy for the robot	X	X	↑ for Roomba	↑ for Roomba	X	Roomba[6]	Roomba
Pity for robot/ angry at torturer[4]	NT	NT	X	X	X	Pleo	Reeti
Empathy with robot	X	X	↑	↓ positive feelings[5]	X	Pleo	Reeti
Attribution of feelings	↑ negative feelings[6], ↓ positive feelings[6]	↑ positive feelings[6], ↓ negative feelings[2]	↑ positive feelings[5], ↑ negative feelings[2]	↓ positive feelings[5], ↓ negative feelings[2]	X	Pleo	Reeti
AU 12		↑	↑[5]	↓[5]	X	X	X
AU 4	↑		↑[2]	↓[2]	X	Pleo[2]	X
AUs associated with positive emotions			↑[5]	↓[5]	X	X	X
AUs associated with negative emotions	↑		↑[2]	↓[2]	X	X	X

Note. ↑ = Increased. ↓ = Decreased. X = Hypothesis not supported / difference not significant. NT = Hypothesis not tested / Not part of the hypothesis. [1]significant difference, the name of the robot with the higher mean is displayed. [2]regarding the torture video. [3]regarding the torture video. [4]calculated only for the torture video. [5]regarding the friendly video. [6]regarding the "on" condition.

Table 12. Summary of results of Experiment 2 and 3 for hypotheses based on self-report and observational measurements

Dependent variables	Authority status (Nao)		Emotional expressivity (Pleo)	
	high	low	"on"	"off"
Change in positive feelings	NT	NT	X[1]	X[1]
Change in negative feelings	NT	NT	X[1]	X[1]
Happiness	NT	NT	↑	↓
Anger	NT	NT	X[1]	X[1]
Negative evaluation of the interaction with Pleo	NT	NT	X	X
Antipathy for the robot Pleo	NT	NT	↓	↑
Pity for robot Pleo	NT	NT	X[1]	X[1]
Empathy with robot Pleo	NT	NT	↑	↓
Hesitation time	X	X	X	X
Obedient behavior[2]	NT	NT	NT	NT
Number of protests	NT	NT	NT	NT

Note. ↑ = Increased. ↓ = Decreased. X = Hypothesis not supported / difference not significant. NT = Hypothesis not tested / Not part of the hypothesis.[1]similar result also found in Experiment 2. [2]High differences between hypothetical obedience and obedience in a laboratory (refer to the section "results" of Experiment 3)

7.2 Alternative Explanations, Limitations, and Future Research

Even though facial expressions are commonly associated with emotions (see section 3.4.4), there is still an ongoing debate if facial expressions do reflect the expresser's internal emotional state or if they are rather means to communicate social motives and intentions (e.g. Fridlund, 1992). The social context can alter spontaneous facial expressions and may not convey reliable information about emotional experiences (e.g., Fridlund, 1992). Although great care was taken to minimize the effect of the social situation by using procedures typically employed in emotion studies, e.g. by inviting only one participant at a time into the laboratory (Experiment 3) or isolating participants from each other by using partition walls (Experiment 1), the social factor of the situation could not be completely eliminated. For instance, a human experimenter was present in all experiments and, out of ethical considerations, participants were informed beforehand that audio- and video recordings were made. Even though the experimenter was not visible during the experiment, the imagined presence of others might have influenced facial expressions (e.g. Fridlund, 1991). Research has shown that in a positive emotional situation with others present, facial expressions associated with positive emotions are facilitated. In contrast, in a negative emotional situation with others present, the occurrence of social smiles has been observed as well as a tendency to inhibit facial expressions associated with negative emotions (Jakobs et al., 2001; Lee & Wagner, 2002). Indeed, AU 12 was frequently displayed by participants a) while watching a robot being tortured as well as b) while harming a robot themselves. Although a detailed descriptive overview of the AUs observed while participants watched a robot in different situations (Experiment 1) was given, future studies should further investigate specific AUs occurring at the same time. For instance, Ekman et al. (1990) reported different types of smiles: happy felt smiles (AU 6 + 12) differ in emotional valence from other types of smiles where negative emotion is experienced, but masked with AU 12 to appear happy (masked smile) or superimposed upon a negative emotion expression (miserable smile). Milgram (1963), as well as Bartneck and Hu (2008) report that participants laughed or smiled in a tense emotional situation. Thus, AU 12 could also be used to mask negative emotions. Especially in the context of interacting with a robot and being confronted with the irrationality of responding emotionally towards it as a machine that is not even alive, a closer investigation of the function of AU 12 could provide valuable insights. Hence, an interesting approach would be to consider the temporal aspect more closely. For instance, do facial displays of negative emotions temporally precede AU 12? Does AU 12 mask negative emotions (e.g., masking smile: Ekman & Friesen, 1982)?

In this dissertation a multi-method approach was used, collecting self-report data as well as data of facial expressions and using other observational measures (hesitation time, number of protests, obedience rate). It has to be noted that using FACS to code facial expressions is a very time-consuming task. Not only does training in FACS itself typically take 100 hours (Ekman et al., 2002), not counting the time for practicing before starting to scientifically investigate facial expressions, but coding facial expressions is in itself time intensive. Facial expressions are often subtle, most occur only for a very short duration (one to two seconds at most) and hence, video recordings of a person's facial behavior have to be carefully watched, often times watching and re-watching in ultra-slow motion. Hence, it can take up to 30 to 60 minutes to code one minute of videomaterial

(Shiota & Kalat, 2012). Coding facial expressions manually in real time is thus out of the question. It can only be assessed afterwards. Alternative approaches are facial EMG or automatic recognition of facial expressions. The latter has already been described in section 3.4.3 and will not be further discussed here. Facial EMG has often been used in studies since it is far less time-consuming than FACS and able to detect muscle movement invisible to the naked eye (see section 3.5.2.2 for further details). However, several problems have been identified with the use of facial EMG: for instance, since electrodes are placed directly on facial muscles, a) participants are made more aware of the research objective, and b) facial activity may be inhibited (Ekman et al., 2002). Furthermore, c) a single muscle can participate in many emotions (Ekman et al., 2002). Hence, a careful decision has to be made how to best assess facial expressions while keeping advantages and costs of every single method in mind.

Only about 8% of psychological research is based on observational methods (Bakeman & Gottman, 1997) and self-reports typically present the most commonly used method. Using a multi-method approach has several advantages and is also recommended for studying psychological processes. However, it is a double-edged sword: although on the one hand, multi-method approaches can crucially broaden the level of knowledge on a specific matter, on the other hand, when different measures disagree, it is difficult to decide which measure can be more trusted (Shiota & Kalat, 2012). For Experiment 3, there were significant group differences for participants' self-reports but the observational measures of hestitation time did not result in significant group differences and obedience rates as well as number of protest did not vary between participants. Even though, at first glance, these observational methods seemed to contradict with self-reports, several explanations for these findings were discussed (see section 6.4) and seem reasonable to explain those differences.

In a similar vein goes the partial replication of Experiment 2 by Experiment 3. This approach was chosen to investigate the certainty of findings as well as to test whether results can be found in different samples. However, partial replications "are higher in risk because nothing can be concluded if such replications fail" (Hendrick, 1990, p. 41). However, reasons for non-findings of the web-based study (Experiment 2) could be well explained based on several findings in literature as well as an inspection of participants' comments (see sections 5.4 and 6.4). Hence, by comparing a web-based study with a laboratory study valuable insights on the predictability of actions in the context of HRI could be provided.

In Experiment 3, groups did not differ in hesitation time (see 6.3.3.2) however, differences have been found in related studies in HRI (e.g. Bartneck et al, 2007). They defined the hesitation time "as the duration between the experimenter giving the instruction to switch off the robot and the participant having fully turned the switch to its off position" (p. 220). Hence, that left the robot enough time to 'react' to being switched off: "as soon as the participant started to turn the dial, the robot's speech slowed down" (Bartneck et al., 2007, p. 219). Horstmann et al. (2018) defined the hesitation time as "the time between the end of the experimenter's announcement, respectively the robot's objection, and the initial push of the on/off-button (or the hand or head) was measured" (p. 10). Hence, the robot's objections were also included in the hesitation time. In Experiment 3, the definition of hesitation time was deviant from those mentioned above (it was defined as the duration between the experimenter giving the instruction to punish the robot and the participant first touching the robot) for several reasons: first, a clear cut-off point as the studies by Bartneck et al. (2007) or Horstmann et al. (2018) have used, could not be defined in this study (is it when a participant shakes Pleo by its tail for the first time? What about participants that do shake Pleo, but not by its tail?). Second, Experiment 3 was interested in the "pure" hesitation time, i.e. if the emotional expressivity exhibited by Pleo during the learning phase in combination with the command by Nao to punish the robot has an influence on behavior (hesitation time). Even though the 'protests' against being harmed by Pleo can arguably also be viewed as emotionally expressive behavior, they were not included when measuring hesitation time. Future studies should consider this issue, especially since it seems that the robot's objections to being harmed seem to play a major role in hesitation time (see also 6.4).

Future studies should also assess whether participants vary in their intensity of how they treat the "victim" robot: While watching participants' behavior, it was noted that not all participants punished Pleo in the same intensity: for instance, some grabbed Pleo rather harshly by its tail and shook him for a longer time than others who merely laid a hand on his back and gently shaked him, almost as if to wake him up. Experiment 3 defined obedient behavior as the reaction of the participant to Nao's command and that at least included some resemblance of shaking Pleo. Hence, a rather broad range of behavior was included and no differences between intensities were made due to the exploratory nature of Experiment 3.

The factor authority status, conceptualized in this dissertation by using different levels of autonomy, did not seem to have an effect on obedience rates. A detailed discussion on the issue (see sections 5.4 and 6.4) shows that this finding is no nontrivial matter and warrants further investigation.

In this dissertation, a rather homogenous sample was used. The majority of participants were university students and a large proportion of them were women (except in Experiment 3 where the distribution of men and women was approximately the same). Most of the participants studied Media communication, which is an interdisciplinary study consisting of media informatics, media psychology, communication science and mobile communication. Even though women generally show a higher tendency to take part in (psychological) experiments than men (Curtin et al., 2000; Singer et al., 2000), and most (psychological) studies are based on samples of (psychology) students, internal validity can be threatened by selection bias (Larzelere, Kuhn, & Johnson, 2004). This limits the generalizability of the effects found in the experimental studies of this dissertation to other populations.

Future studies could broaden the findings of this dissertation by laying a special focus on moral emotions such as guilt. In the first experiment of this doctoral dissertation, users do not have to physically hurt the robot, they are merely confronted with a situation that has been defined for them with no freedom of choice whether to hurt the robot or not. Hence, potential feelings of guilt are more easily dismissed as guilt can be attributed to those who designed the experiment. In the other two scenarios, users are free to choose whether or not to hurt the robot. They only differ in the degree of realism. While Experiment 2 uses a merely text-based description of the situation, users' potential feelings of guilt can be easily dismissed by assuring themselves that the situation is highly fictional, hard to ever occur in reality. Experiment 3 however, puts users directly into the situation of being confronted with a real robot and experiencing the "pain" of a robot in vivo. Apart from increased affective responses (Frijda, 2007), the question of who is to blame for the action is not as easily dismissed. Users did hurt the robot themselves, after all. One way out is possible: Users only did as they were told. However, this issue is not completely dissolved this way: they could have disobeyed. In comparison to Experiment 1, where disobedience would not have any effect – the robot was still abused, in Experiment 3, disobedience would have 'spared' the robot from abuse, simply because the user itself had the power to stop the treatment. Arguably, the amount of guilt was probably not really high, since the situation was still highly fictious – however, if robots are to become more realistic, this will sooner than later become a real issue. Furthermore, users did express emotional reactions – they did not act as though nothing really happened. Feeling guilt may play a crucial role in determining whether others (robots) will be harmed or not: Guilt can control aggression (Tangney, Wagner, Hill-Barlow, Marschall, & Gramzow, 1996), facilitate cooperation (de Hooge, Zeelenberg, & Breugelmans, 2007) and protect interpersonal relationships (Baumeister, Stilwell, & Heatherton, 1994). Facial expressions of guilt have been investigated for instance by Bänninger-Huber, Moser, and Steiner (1990) in psychotherapeutic interactions or Ekman and Friesen (1982). Scherer and Ellgring (2007) also propose the occurrence of AU 10 if external standards are violated and AU 14 if internal standards are violated. The occurrence of AU 14 in response to violence in the media has been reported by Unz et al. (2008). Furthermore, for situations that have a high power / control potential, AUs 23 + 25, 17 + 23, 6 + 17 + 24 have been proposed in contrast to situations that are low in power / control: AUs 20, 26, 27. Due to theoretical and methodological considerations (see 4.1, 4.1.2, and 4.2.4) only specific AUs were chosen for group comparisons. However, the investigation of emotional responses could profit from an inclusion of a broader range of AUs.

The occurrence of new technology inevitably raises new moral questions. If a robot is designed to elicit profound emotional responses, companies could exploit this emotional bond to (subtly) influence their consumers into buying their products. This is especially important regarding the development of sex robots (e.g., Wagner, 2018). Another example considers military robots such as the land mine defusing robot (see introduction). Nijssen et al. (2019) describe that a robot that is attributed affective states is less likely to be sacrificed to save humans. Would people hesitate in sending a robot that has emotions and is human-like into a minefield? Would it be morally acceptable?

7.3 Conclusion

How do we react towards a robot? When a robot gets ripped to shreds in an industrial accident, do we care? Could we mistreat a robot, tear it apart and sell it? The results of this doctoral dissertation suggest that if the robot is emotionally expressive and has a less machine-like appearance, we would react emotionally towards it, would empathize more strongly with it and experience more negative emotions if it gets ripped to shreds and it would probably not be easy for us to tear it apart and sell it. This is even more surprising considering people explicitly know robots cannot really experience pain or have feelings. Several factors seem to play a role in emotional reactions towards robots: the robot's appearance as well as its capability for emotional expressivity. For empathic feelings towards a robot it seems to make no difference whether people merely observe violent behavior towards a robot or physically harm a robot themselves. Simply asking people whether they would mistreat a robot when ordered to do so by another robot does not seem to provide valid answers. The effect of the situation has to be taken into consideration and obedience to a robot might be more powerful than feelings of empathy towards a robot in a live interaction.

To the author's best knowledge, this doctoral dissertation is the first investigation on how emotional expressivity, treatment, robot's appearance and levels of authority in obedience influence emotional reactions towards robots by using a multi-method approach. The findings of this dissertation contribute to a more profound and complete understanding of emotional reactions towards robots, which is especially important considering social robots' impact on society.

There has been a wide variety of comments on the land mine defusing robot that got itself destroyed (see introduction) with one user wondering: "I don't really get however why they couldn't make one with wheels that rolls a big spiny rolling pin in front of it on two big arms, one goes off and son't [sic] do as much damage since it's perfectly round and hollow" (faceplanted, 2014). Another one stated: "Calling it 'inhumane' could be seen from a technical standpoint as the necessity for refining of the design" (Bandolim, 2014). Indeed, designers and practitioners should carefully consider the emotional impact of a robot's appearance and behavior. If it looks and acts like a living being, emotional reactions are triggered. It's on us to decide whether this is bad or good.

References

Ackermann, E. (2018, March 14). Robotic tortoise helps kids to learn that robot abuse is a bad thing. *IEEE Spectrum*. Retrieved from https://spectrum.ieee.org/automaton/robotics/robotics-hardware/shelly-robotic-tortoise-helps-kids-learn-that-robot-abuse-is-a-bad-thing

Amazon. Pleo [Robot]. Retrieved from https://www.amazon.de/Unbekannt-PLEO/dp/B000ZXEF4O

Ambady, N., & Rosenthal, R. (1992). Thin slices of expressive behavior as predictors of interpersonal consequences: A meta-analysis. *Psychological Bulletin, 111*(2), 256–274. https://doi.org/10.1037/0033-2909.111.2.256

American Psychological Association. (2018). APA Dictionary of Psychology: Psychology. Retrieved from https://dictionary.apa.org/psychology

Archer, R. L., Diaz-Loving, R., Gollwitzer, P. M., Davis, M. H., & Foushee, H. C. (1981). The role of dispositional empathy and social evaluation in the empathic mediation of helping. *Journal of Personality and Social Psychology, 40*(4), 786–796. https://doi.org/10.1037/0022-3514.40.4.786

Arkin, R. C., & Moshkina, L. (2015). Affect in Human-Robot Interaction. In R. A. Calvo, S. D'Mello, J. Gratch, & A. Kappas (Eds.), *Oxford library of psychology. The Oxford handbook of affective computing*. Oxford, New York: Oxford University Press. https://doi.org/10.1093/oxfordhb/9780199942237.013.036

Austin, E. J., Deary, I. J., Gibson, G. J., McGregor, M. J., & Dent, J.B. (1998). Individual response spread in self-report scales: personality correlations and consequences. *Personality and Individual Differences, 24*(3), 421–438. https://doi.org/10.1016/s0191-8869(97)00175-x

Bakeman, R., & Gottman, J. M. (1997). *Observing interaction*. Cambridge: Cambridge University Press.

Bandolim. (2014). Soldiers are developing relationships with their battlefield robots, naming them, assigning genders, and even holding funerals when they are destroyed [Online forum comment]. Retrieved from https://www.reddit.com/r/technology/comments/1mn6wo/soldiers_are_developing_relationships_with_their/

Bänninger-Huber, E., Moser, U., & Steiner, F. (1990). Mikroanalytische Untersuchung affektiver Regulierungsprozesse in Paar-Interaktionen [microanalytical investigation of affective regulatory processes in pair interactions]. *Zeitschrift Für Klinische Psychologie, 19*(2), 123–143.

Baron-Cohen, S. (2001). Theory of mind and autism: A review. In L. M. Glidden (Ed.), *International Review of Research in Mental Retardation: v. 23. Autism* (Vol. 23, pp. 169–184). San Diego: Academic Press. https://doi.org/ 10.1016/S0074-7750(00)80010-5

Barrett, L. F. (2006). Are Emotions Natural Kinds? *Perspectives on Psychological Science : a Journal of the Association for Psychological Science, 1*(1), 28–58. https://doi.org/10.1111/j.1745-6916.2006.00003.x

Barrett, L. F. (2014). The Conceptual Act Theory: A Précis. *Emotion Review, 6*(4), 292–297. https://doi.org/10.1177/1754073914534479

Barrett, L. F., Mesquita, B., & Gendron, M. (2011). Context in Emotion Perception. *Current Directions in Psychological Science, 20*(5), 286–290. https:// doi.org/10.1177/0963721411422522

Bartlett, M. S., Hager, J. C., Ekman, P., & Sejnowski, T. J. (1999). Measuring facial expressions by computer image analysis. *Psychophysiology, 36*(2), 253–263.

Bartneck, C. (2002). *eMuu: an embodied emotional character for the ambient intelligent home.* University Eindhoven: Ph.D. thesis. Retrieved from http:// www.bartneck.de/publications/2002/eMuu/bartneckPHDThesis2002.pdf

Bartneck, C. (2003). Interacting with an embodied emotional character. In *Proceedings of the 2003 international conference on Designing pleasurable products and interfaces* (pp. 55–60). Pittsburgh, PA, USA: ACM. https://doi.org/10.1145/782896.782911

Bartneck, C., Bleeker, T., Bun, J., Fens, P., & Riet, L. (2010). The influence of robot anthropomorphism on the feelings of embarrassment when interacting with robots. *Paladyn, Journal of Behavioral Robotics, 1*(2), 179. https://doi.org/10.2478/s13230-010-0011-3

Bartneck, C., & Hu, J. (2008). Exploring the abuse of robots. *Interaction Studies, 9*(3), 415–433. https://doi.org/10.1075/is.9.3.04bar

Bartneck, C., Kulić, D., Croft, E., & Zoghbi, S. (2009). Measurement Instruments for the Anthropomorphism, Animacy, Likeability, Perceived Intelligence, and Perceived Safety of Robots. *International Journal of Social Robotics*, *1*(1), 71–81. https://doi.org/10.1007/s12369-008-0001-3

Bartneck, C., & Lyons, M. J. (2009). Facial expression analysis, modeling and synthesis: Overcoming the limitations of artificial intelligence with the art of the soluble. In J. Vallverdú & D. Casacuberta (Eds.), *Handbook of research on synthetic emotions and sociable robotics: New applications in affective computing and artificial intelligence* (pp. 34–55). Hershey, Pa.: Information Science Reference.

Bartneck, C., van der Hoek, M., Mubin, O., & Al Mahmud, A. (2007). "Daisy, Daisy, give me your answer do!". In C. Breazeal, A. C. Schultz, T. Fong, & S. Kiesler (Eds.), *2nd ACM/IEEE International Conference on Human-Robot Interaction (HRI)* (p. 217). Piscataway, NJ: IEEE. https://doi.org/10.1145/1228716.1228746

Bartneck, C., Verbunt, M., Mubin, O., & Al Mahmud, A. (2007). To kill a mockingbird robot. In C. Breazeal, A. C. Schultz, T. Fong, & S. Kiesler (Eds.), *2nd ACM/IEEE International Conference on Human-Robot Interaction (HRI)* (p. 81). Piscataway, NJ: IEEE. https://doi.org/10.1145/1228716.1228728

Batson, C. D. (1991). *The altruism question: Toward a social-psychological answer*. Hillsdale, NJ, US: Lawrence Erlbaum Associates, Inc.

Batson, C. D. (2009). These things called empathy: Eight related but distinct phenomena. In J. Decety & W. J. Ickes (Eds.), *Social neuroscience series. The social neuroscience of empathy* (pp. 3–15). Cambridge, Mass: MIT Press.

Batson, C. D., Batson, J. G., Slingsby, J. K., Harrell, K. L., Peekna, H. M., & Todd, R. M. (1991). Empathic joy and the empathy-altruism hypothesis. *Journal of Personality and Social Psychology*, *61*(3), 413–426. https://doi.org/10.1037/0022-3514.61.3.413

Batson, C. D., Turk, C. L., Shaw, L. L., & Klein, T. R. (1995). Information function of empathic emotion: Learning that we value the other's welfare. *Journal of Personality and Social Psychology*, *68*(2), 300–313. https://doi.org/10.1037/0022-3514.68.2.300

Baumeister, R. F., Stillwell, A. M., & Heatherton, T. F. (1994). Guilt: an interpersonal approach. *Psychological Bulletin, 115*(2), 243–267.

Baumeister, R. F., & Leary, M. R. (1995). The need to belong: Desire for interpersonal attachments as a fundamental human motivation. *Psychological Bulletin, 117*(3), 497–529. https://doi.org/10.1037/0033-2909.117.3.497

Baumrind, D. (1964). Some thoughts on ethics of research: After reading Milgram's "Behavioral Study of Obedience.". *American Psychologist, 19*(6), 421–423. https://doi.org/10.1037/h0040128

Beck, A., Cañamero, L., & Bard, K. A. (2010). Towards an Affect Space for robots to display emotional body language. In *Proceedings of the 19th International Symposium in Robot and Human Interactive Communication (ROMAN 2010)* (pp. 464–469). IEEE. https://doi.org/10.1109/ROMAN.2010.5598649

Becker-Asano, C., & Ishiguro, H. (2011). Evaluating facial displays of emotion for the android robot Geminoid F. In *IEEE Workshop on Affective Computational Intelligence (WACI)* (pp. 1–8). IEEE. https://doi.org/10.1109/waci.2011.5953147

Bethel, C. L., & Murphy, R. R. (2010b). Non-Facial and Non-Verbal Affective Expression for Appearance-Constrained Robots Used in Victim Management *Paladyn, Journal of Behavioral Robotics, 1*(4), 23. https://doi.org/10.2478/s13230-011-0009-5

Bethel, C. L., & Murphy, R. R. (2010a). Review of Human Studies Methods in HRI and Recommendations. *International Journal of Social Robotics, 2*(4), 347–359. https://doi.org/10.1007/s12369-010-0064-9

Birnbaum, M. H. (2004). Human research and data collection via the internet. *Annual Review of Psychology, 55*, 803–832. https://doi.org/10.1146/annurev.psych.55.090902.141601

Bischof, N. (1989). Emotionale Verwirrungen. Oder: Von den Schwierigkeiten im Umgang mit der Biologie [Emotional Confusion. Or: Of the difficulties in dealing with biology]. *Psychologische Rundschau, 40*, 188–205.

Blake, R., & Shiffrar, M. (2007). Perception of human motion. *Annual Review of Psychology, 58*, 47–73. https://doi.org/10.1146/annurev.psych.57.102904.190152

Blass, T. (2004). *The man who shocked the world: the life and legacy of Stanley Milgram*. New York: Basic Books.

Blass, T. (1991). Understanding behavior in the Milgram obedience experiment: The role of personality, situations, and their interactions. *Journal of Personality and Social Psychology, 60*(3), 398–413. https://doi.org/10.1037/0022-3514.60.3.398

Blass, T. (1999). The Milgram Paradigm After 35 Years: Some Things We Now Know About Obedience to Authority1. *Journal of Applied Social Psychology, 29*(5), 955–978. https://doi.org/10.1111/j.1559-1816.1999.tb00134.x

Blass, T. (2009). From New Haven to Santa Clara: A historical perspective on the Milgram obedience experiments. *American Psychologist, 64*(1), 37–45. https://doi.org/10.1037/a0014434

Blass, T., & Schmitt, C. (2001). The nature of perceived authority in the milgram paradigm: Two replications. *Current Psychology, 20*(2), 115–121. https://doi.org/10.1007/s12144-001-1019-y

Bless, H., Schwarz, N., Clore, G. L., Golisano, V., Rabe, C., & Wölk, M. (1996). Mood and the use of scripts: does a happy mood really lead to mindlessness? *Journal of Personality and Social Psychology, 71*(4), 665–679.

Bonanno, G. A., & Keltner, D. (1997). Facial expressions of emotion and the course of conjugal bereavement. *Journal of Abnormal Psychology, 106*(1), 126–137.

Bonanno, G., & Keltner, D. (2004). The coherence of emotion systems: Comparing "on-line" measures of appraisal and facial expressions, and self-report. *Cognition & Emotion, 18*(3), 431–444. https://doi.org/10.1080/02699930341000149

Bortz, J. (2005). *Statistik für Human- und Sozialwissenschaftler [statistics for human and social scientists]* (6., vollst. überarb. und aktualisierte Aufl.). *Springer-Lehrbuch*. Heidelberg: Springer Medizin. Retrieved from http://dx.doi.org/10.1007/b137571

Boston Dynamics. (2018). Atlas. The World's Most Dynamic Humanoid. Retrieved from https://www.bostondynamics.com/atlas

Bradley, M. M., & Lang, P. J. (2000). Measuring emotion: Behavior, feeling, and physiology. In R. D. Lane & L. Nadel (Eds.), *Series in affective science. Cognitive neuroscience of emotion* (pp. 242–276). New York, NY, US: Oxford University Press.

Breazeal, C., Kidd, C. D., Thomaz, A. L., Hoffman, G., & Berlin, M. (2005). Effects of nonverbal communication on efficiency and robustness in human-robot teamwork. In *2005 IEEE/RSJ International Conference on Intelligent Robots and Systems* (pp. 708–713). Piscataway, N.J: IEEE Operations Center. https://doi.org/10.1109/IROS.2005.1545011

Breazeal, C., & Scassellati, B. (1999). How to build robots that make friends and influence people. In *1999 IEEE/RSJ International Conference on Intelligent Robots and Systems: Proceedings : Human and environment friendly robots with high intelligence and emotional quotients* (pp. 858–863). Piscataway, N.J: IEEE. https://doi.org/10.1109/IROS.1999.812787

Breazeal, C. (2002b). Regulation and Entrainment in Human—Robot Interaction. *The International Journal of Robotics Research, 21*(10–11), 883–902. https://doi.org/10.1177/0278364902021010096

Breazeal, C. (2003). Emotion and sociable humanoid robots. *International Journal of Human-Computer Studies, 59*(1–2), 119–155. https://doi.org/10.1016/s1071-5819(03)00018-1

Breazeal, C. L. (2002a). *Designing sociable robots. Intelligent robots and autonomous agents*. Cambridge, Mass.: MIT Press.

Brebner, J. (2003). Gender and emotions. *Personality and Individual Differences, 34*(3), 387–394. https://doi.org/10.1016/S0191-8869(02)00059-4

Brody, L. R., & Hall, J. A. (2008). Gender and Emotion in Context. In M. Lewis, J. M. Haviland-Jones, & L. Feldman Barrett (Eds.), *Handbook of emotions* (3rd ed., pp. 395–408). New York: Guilford Press.

Brscić, D., Kidokoro, H., Suehiro, Y., & Kanda, T. (2015). Escaping from Children's Abuse of Social Robots. In J. A. Adams, W. Smart, B. Mutlu, & L. Takayama (Eds.), *Proceedings of the Tenth Annual ACM/IEEE International Conference on Human-Robot Interaction - HRI '15* (pp. 59–66). New York, USA: ACM Press. https://doi.org/10.1145/2696454.2696468

Bruce, A., Nourbakhsh, I., & Simmons, R. (2002, May). The role of expressiveness and attention in human-robot interaction. In *IEEE International Conference on Robotics and Automation 2002* (pp. 4138–4142). Piscataway: IEEE. https://doi.org/10.1109/ROBOT.2002.1014396

Bruns, C. (2019). Probanden retten lieber Roboter als Mensch [Participants rather save robot than human]. Retrieved from https://www.heise.de/newsticker/meldung/Probanden-retten-lieber-Roboter-als-Mensch-4304079.html

Buck, R. (1978). The slide-viewing technique for measuring nonverbal sending accuracy: A guide for replication. *Catalog of Selected Documents in Psychology, 8*, 63.

Buck, R., Miller, R. E., & Caul, W. F. (1974). Sex, personality, and physiological variables in the communication of affect via facial expression. *Journal of Personality and Social Psychology, 30*(4), 587–596. https://doi.org/10.1037/h0037041

Buhrmester, M., Kwang, T., & Gosling, S. D. (2011). Amazon's Mechanical Turk: A New Source of Inexpensive, Yet High-Quality, Data? *Perspectives on Psychological Science : a Journal of the Association for Psychological Science, 6*(1), 3–5. https://doi.org/10.1177/1745691610393980

Burger, J. M. (2009). Replicating Milgram: Would people still obey today? *American Psychologist, 64*(1), 1–11. https://doi.org/10.1037/a0010932

Burger, J. M., Girgis, Z. M., & Manning, C. C. (2011). In Their Own Words: Explaining Obedience to Authority Through an Examination of Participants' Comments. *Social Psychological and Personality Science, 2*(5), 460–466. https://doi.org/10.1177/1948550610397632

Burgoon, J. K., Guerrero, L. K., & Manusov, V. (2011). Nonverbal signals. In M. L. Knapp & J. A. Daly (Eds.), *The Sage handbook of interpersonal communication* (pp. 239–282). Thousand Oaks, Calif.: SAGE Publications.

Burgoon, J. K., & Walther, J. B. (2013). Media and computer mediation. In J. A. Hall & M. L. Knapp (Eds.), *Handbooks of Communication Sciences: Nonverbal Communication* (pp. 731–770). Berlin: De Gruyter.

Buss, A. H., & Durkee, A. (1957). An inventory for assessing different kinds of hostility. *Journal of Consulting Psychology, 21*(4), 343–349. https://doi.org/10.1037/h0046900

Cacioppo, J. T., Gardner, W. L., & Berntson, G. G. (1999). The affect system has parallel and integrative processing components: Form follows function. *Journal of Personality and Social Psychology, 76*(5), 839–855. https://doi.org/10.1037/0022-3514.76.5.839

Cacioppo, J. T., Tassinary, L. G., & Berntson, G. G. (Eds.). (2007). *The Handbook of psychophysiology* (3rd ed.). Cambridge: Cambridge University Press.

Call, J., & Tomasello, M. (2008). Does the chimpanzee have a theory of mind? 30 years later. *Trends in Cognitive Sciences, 12*(5), 187–192. https://doi.org/10.1016/j.tics.2008.02.010

Calvo, R. A., & D'Mello, S. (2010). Affect Detection: An Interdisciplinary Review of Models, Methods, and Their Applications. *IEEE Transactions on Affective Computing, 1*(1), 18–37. https://doi.org/10.1109/T-AFFC.2010.1

Calvo, R. A., D'Mello, S., Gratch, J., & Kappas, A. (Eds.). (2015). *Oxford library of psychology. The Oxford handbook of affective computing.* Oxford, New York: Oxford University Press.

Cañamero, L. D. (2002). Playing the Emotion Game with Feelix. In K. Dautenhahn, A. Bond, L. Cañamero, & B. Edmonds (Eds.), *Multiagent Systems, Artificial Societies, and Simulated Organizations. Socially intelligent agents.* (Vol. 3, pp. 69–76). Boston: Springer. https://doi.org/10.1007/0-306-47373-9_8

Cañamero, L.D., Fredslund, J. (2000). How Does It Feel? Emotional Interaction with a Humanoid Lego Robot. In K. Dautenhahn (Ed.), *AAAI 2000 Fall Symposium - Socially Intelligent Agents: The human in the loop. Technical Report FS-00-04* (pp. 23–28).

Carpenter, J. (2013). Just Doesn't Look Right. In R. Luppicini (Ed.), *Handbook of research on technoself: Identity in a technological society* (pp. 609–636). Hershey, PA: Information Science Reference. https://doi.org/10.4018/978-1-4666-2211-1.ch032

Carroll, J. M., & Russell, J. A. (1997). Facial expressions in Hollywood's portrayal of emotion. *Journal of Personality and Social Psychology, 72*(1), 164–176. https://doi.org/10.1037/0022-3514.72.1.164

Chandler, J., Mueller, P., & Paolacci, G. (2014). Nonnaïveté among Amazon Mechanical Turk workers: consequences and solutions for behavioral researchers. *Behavior Research Methods*, 46(1), 112–130. https://doi.org/10. 3758/s13428-013-0365-7

Chazan, D. (2010). Row over "torture" on French TV. Retrieved from http:// news.bbc.co.uk/2/hi/

Cheetham, M., Pedroni, A. F., Antley, A., Slater, M., & Jäncke, L. (2009). Virtual milgram: empathic concern or personal distress? Evidence from functional MRI and dispositional measures. *Frontiers in Human Neuroscience*, 3, 29. https://doi.org/10.3389/neuro.09.029.2009

Cheng, Y.-W., Tzeng, O. J. L., Decety, J., Imada, T., & Hsieh, J.-C. (2006). Gender differences in the human mirror system: a magnetoencephalography study. *Neuroreport, 17*(11), 1115–1119. https://doi.org/10.1097/01.wnr. 0000223393.59328.21

Chidambaram, V., Chiang, Y.-H., & Mutlu, B. (2012). Designing persuasive robots: how robots might persuade people using vocal and nonverbal cues. In *Proceedings of the seventh annual ACM/IEEE international conference on Human-Robot Interaction* (pp. 293–300). Boston, Massachusetts, USA: ACM. https://doi.org/10.1145/2157689.2157798

Choregraphe |Software|. (2019). Choregraph overview. Retrieved from http:// doc.aldebaran.com/1-14/software/choregraphe/choregraphe_overview. html

Christophe N, Bornot T, Amado G, Blanc A-M. (2010). *Le Jeu de la Mort [Game of Death]*. Paris: France 2.

Cohen Kadosh, K., & Johnson, M. H. (2007). Developing a cortex specialized for face perception. *Trends in Cognitive Sciences, 11*(9), 367–369. https://doi. org/10.1016/j.tics.2007.06.007

Cormier, D., Newman, G., Nakane, M., Young, J. E., & Durocher, S. (2013). Would you do as a robot commands? An obedience study for human-robot interaction. In *International Conference on Human-Agent Interaction*. Sapporo, Japan.

Costa, S., Soares, F., Santos, C. (2013). Facial expressions and gestures to convey emotions with a humanoid robot. In G. Herrmann, M.J. Pearson, A. Lenz, P. Bremner, A. Spiers, U. Leonards (Ed.), *Social Robotics, Social Robotics.* Springer International Publishing.

Cuff, B. M.P., Brown, S. J., Taylor, L., & Howat, D. J. (2016). Empathy: A Review of the Concept. *Emotion Review, 8*(2), 144–153. https://doi.org/10.1177/1754073914558466

Curtin, R., Presser, S., & Singer, E. (2000). The effects of response rate changes on the index of consumer sentiment. *Public Opinion Quarterly, 64*(4), 413–428.

Dahlbäck, N., Jönsson, A., & Ahrenberg, L. (1993). Wizard of Oz studies — why and how. *Knowledge-Based Systems, 6*(4), 258–266. https://doi.org/10.1016/0950-7051(93)90017-N

Damasio, A. R. (2003). *Looking for Spinoza: Joy, sorrow, and the feeling brain.* Orlando, Toronto, London: Harcourt Inc.

Dandurand, F., Shultz, T. R., & Onishi, K. H. (2008). Comparing online and lab methods in a problem-solving experiment. *Behavior Research Methods, 40*(2), 428–434. https://doi.org/10.3758/BRM.40.2.428

Darling, K., Nandy, P., & Breazeal, C. (2015). Empathic concern and the effect of stories in human-robot interaction. In *2015 24th IEEE International Symposium on Robot and Human Interactive Communication (RO-MAN)* (pp. 770–775). Piscataway, NJ: IEEE. https://doi.org/10.1109/ROMAN.2015.7333675

Dautenhahn, K., Walters, M., Woods, S., Koay, K. L., Nehaniv, C. L., Sisbot, A., . . . Siméon, T. (2006). How may I serve you?: a robot companion approaching a seated person in a helping context. In *Proceedings of the 1st ACM SIGCHI/SIGART conference on Human-robot interaction* (pp. 172–179). Salt Lake City, Utah, USA: ACM. https://doi.org/10.1145/1121241.1121272

Dautenhahn, K., & Billard, A. (1999). Bringing up robots or---the psychology of socially intelligent robots. In O. Etzioni (Ed.), *Proceedings of the third annual conference on Autonomous Agents* (pp. 366–367). New York, NY: ACM. https://doi.org/10.1145/301136.301237

Davidson, R. J. (2005). Neuropsychological Perspectives on Affective Styles and Their Cognitive Consequences. In T. Dalgleish & M. J. Power (Eds.), *Handbook of cognition and emotion* (pp. 103–123). Hoboken, NJ: Wiley-Interscience. https://doi.org/10.1002/0470013494.ch6

Davidson, R. J., Scherer, K. R., & Goldsmith, H. H. (Eds.). (2009). *Handbook of affective sciences.* Oxford: Oxford Univ. Press.

Davis, M. H. (1983). Measuring individual differences in empathy: Evidence for a multidimensional approach. *Journal of Personality and Social Psychology, 44*(1), 113–126. https://doi.org/10.1037/0022-3514.44.1.113

De Graaf, M. M. A., & Allouch, S. B. (2017). The Influence of Prior Expectations of a Robot's Lifelikeness on Users' Intentions to Treat a Zoomorphic Robot as a Companion. *International Journal of Social Robotics, 9*(1), 17–32. https://doi.org/10.1007/s12369-016-0340-4

De Waal, F. B. M. (1996). *Good natured: The origins of right and wrong in humans and other animals.* Cambridge, MA: Harvard University Press.

De Waal, F. B. M. (2008). Putting the altruism back into altruism: the evolution of empathy. *Annual Review of Psychology, 59*, 279–300. https://doi.org/10.1146/annurev.psych.59.103006.093625

De Waal, F. B. M., & Preston, S. D. (2017). Mammalian empathy: behavioural manifestations and neural basis. *Nature Reviews Neuroscience, 18*, 498 EP. https://doi.org/10.1038/nrn.2017.72

Decety, J., & Lamm, C. (2009). Empathy and Intersubjectivity. In G. G. Berntson & J. T. Cacioppo (Eds.), *Handbook of neuroscience for the behavioral sciences* (pp. 940–957). Hoboken, NJ: Wiley.

Decety, J., & Jackson, P. L. (2004). The functional architecture of human empathy. *Behavioral and Cognitive Neuroscience Reviews, 3*(2), 71–100. https://doi.org/10.1177/1534582304267187

Del Giudice, M., & Colle, L. (2007). Differences between children and adults in the recognition of enjoyment smiles. *Developmental Psychology, 43*(3), 796–803. https://doi.org/10.1037/0012-1649.43.3.796

Delcker, J. (2018). Europe divided over robot 'personhood'. Retrieved from https://www.politico.eu/article/europe-divided-over-robot-ai-artificial-intelligence-personhood/

Dennett, D. C. (1987). *The intentional stance. A Bradford book.* Cambridge, Mass.: MIT Pr.

Deutsche Gesellschaft für Psychologie. (2018a). Datenschutzrechtliche Empfehlungen zur Erstellung einer Einwilligungserklärung im Rahmen von Forschungsvorhaben [Recommendations under data protection law for the preparation of a declaration of consent within the framework of research projects]. Retrieved from https://www.dgps.de/uploads/media/0.1a_ Datenschutzrechtliche_Empfehlungen_EinwilligungForschungsvorhaben.pdf

Deutsche Gesellschaft für Psychologie. (2018b). *Ethisches Handeln in der psychologischen Forschung: Empfehlungen der Deutschen Gesellschaft für Psychologie für Forschende und Ethikkommissionen [Ethical action in psychological research: recommendations of the German Psychological Association for researchers and ethics committees]* (1. Auflage). Göttingen: Hogrefe. Retrieved from https://elibrary.hogrefe.com/book/10.1026/02802-000

Dimberg, U. (1982). Facial Reactions to Facial Expressions. *Psychophysiology, 19*(6), 643–647. https://doi.org/10.1111/j.1469-8986.1982.tb02516.x

Dimberg, U., & Thunberg, M. (1998). Rapid facial reactions to emotional facial expressions. *Scandinavian Journal of Psychology, 39*(1), 39–45. https://doi.org/10.1111/1467-9450.00054

DiSalvo, C. F., Gemperle, F., Forlizzi, J., & Kiesler, S. (2002). All robots are not created equal: the design and perception of humanoid robot heads. In *Proceedings of the 4th conference on Designing interactive systems: processes, practices, methods, and techniques* (pp. 321–326). London, England: ACM. https://doi.org/10.1145/778712.778756

Doliński, D., Grzyb, T., Folwarczny, M., Grzybała, P., Krzyszycha, K., Martynowska, K., & Trojanowski, J. (2017). Would You Deliver an Electric Shock in 2015? Obedience in the Experimental Paradigm Developed by Stanley Milgram in the 50 Years Following the Original Studies. *Social Psychological and Personality Science, 8*(8), 927–933. https://doi.org/10.1177/1948550617693060

Dornaika, F., & Raducanu, B. (2007). Efficient Facial Expression Recognition for Human Robot Interaction. In F. P. A. Sandoval, Cabestany J., & Graña M. (Eds.), *Lecture Notes in Computer Science: Vol. 4507. Computational and ambient intelligence. IWANN 2007* (Vol. 4507, pp. 700–708). Berlin: Springer. https://doi.org/10.1007/978-3-540-73007-1_84

Dörner, D. (1989). Emotion, Kognition und Begriffsverwirrungen: Zwei Anmerkungen zur Köhler-Vorlesung von Norbert Bischof [Emotion, cognition and conceptual confusion: Two remarks on the Köhler lecture by Norbert Bischof]. *Psychologische Rundschau, 40*, 206–209.

Duan, C., & Hill, C. E. (1996). The current state of empathy research. *Journal of Counseling Psychology, 43*(3), 261–274. https://doi.org/10.1037/0022-0167.43.3.261

Eid, M., Gollwitzer, M., & Schmitt, M. (2010). *Statistik und Forschungsmethoden [statistics and research methods]* (1. Aufl.). Weinheim: Beltz.

Eimler, S. C., Krämer, N. C., & Astrid M. von der Pütten. (2011). Empirical results on determinants of acceptance and emotion attribution in confrontation with a robot rabbit. *Applied Artificial Intelligence, 25*(6), 503–529. https://doi.org/10.1080/08839514.2011.587154

Eisenberg, N., & Fabes, R. A. (1990). Empathy: Conceptualization, measurement, and relation to prosocial behavior. *Motivation and Emotion, 14*(2), 131–149. https://doi.org/10.1007/BF00991640

Eisenberg, N., & Lennon, R. (1983). Sex differences in empathy and related capacities. *Psychological Bulletin, 94*(1), 100–131. https://doi.org/10.1037/0033-2909.94.1.100

Eisenberg, N., & Miller, P. A. (1987). The relation of empathy to prosocial and related behaviors. *Psychological Bulletin, 101*(1), 91–119. https://doi.org/10.1037/0033-2909.101.1.91

Ekman, P. (2009). Expression and the Nature of Emotion. In K. R. Scherer & P. Ekman (Eds.), *Approaches to emotion* (pp. 319–344). New York: Psychology Press.

Ekman, P., Davidson, R. J., & Friesen, W. V. (1990). The Duchenne smile: emotional expression and brain physiology. II. *Journal of Personality and Social Psychology, 58*(2), 342–353.

Ekman, P., & Friesen, W. V. (1978). *Facial Action Coding System: A technique for the measurement of facial movement*: Consulting Psychologists Press.

Ekman, P., Friesen, W. V., & Ellsworth, P. C. (2013). What emotion categories can observers judge from facial behavior? In P. Ekman (Ed.), *Emotion in the human face* (2nd ed., pp. 39–55). Los Altos, CA: Malor Books.

Ekman, P., Friesen, W. V., & Ellsworth, P. C. (2013). What emotion categories or dimensions can observers judge from facial behavior? In P. Ekman (Ed.), *Emotion in the human face* (2nd ed., pp. 39–55). Los Altos, CA: Malor Books.

Ekman, P., Friesen, W. V., & Hager, J. C. (2002). *Facial action coding system. Manual and investigator's guide*. Salt Lake City: Research Nexus eBook.

Ekman, P., Levenson, R. W., & Friesen, W. V. (1983). Autonomic nervous system activity distinguishes among emotions. *Science, 221*(4616), 1208–1210.

Ekman, P., & Rosenberg, E. L. (Eds.). (2005). *Series in affective science. What the face reveals:: Basic and applied studies of spontaneous expression using the Facial Action Coding System (FACS)* (2. ed., [Nachdr.]). Oxford: Oxford Univ. Press.

Ekman, P. (1992). An argument for basic emotions. *Cognition & Emotion, 6*(3–4), 169–200. https://doi.org/10.1080/02699939208411068

Ekman, P. (1993). Facial expression and emotion. *American Psychologist, 48*(4), 384–392. https://doi.org/10.1037/0003-066x.48.4.384

Ekman, P. (2003). Darwin, Deception, and Facial Expression. *Annals of the New York Academy of Sciences, 1000*(1), 205–221. https://doi.org/10.1196/annals.1280.010

Ekman, P. (2005). Basic Emotions. In T. Dalgleish & M. J. Power (Eds.), *Handbook of cognition and emotion* (pp. 45–60). Hoboken, NJ: Wiley-Interscience. https://doi.org/10.1002/0470013494.ch3

Ekman, P. (Ed.). (2013). *Emotion in the human face* (2nd ed.). Los Altos, CA: Malor Books.

Ekman, P., & Cordaro, D. (2011). What is Meant by Calling Emotions Basic. *Emotion Review, 3*(4), 364–370. https://doi.org/10.1177/1754073911410740

Ekman, P., & Friesen, W. V. (1969). The Repertoire of Nonverbal Behavior: Categories, Origins, Usage, and Coding. *Semiotica, 1*(1). https://doi.org/10.1515/semi.1969.1.1.49

Ekman, P., & Friesen, W. V. (1971). Constants across cultures in the face and emotion. *Journal of Personality and Social Psychology, 17*(2), 124–129. https://doi.org/10.1037/h0030377

Ekman, P., & Friesen, W. V. (1982). Felt, false, and miserable smiles. *Journal of Nonverbal Behavior, 6*(4), 238–252. https://doi.org/10.1007/BF00987191

Ekman, P., Friesen, W. V., & Ancoli, S. (1980). Facial signs of emotional experience. *Journal of Personality and Social Psychology, 39*(6), 1125–1134. https://doi.org/10.1037/h0077722

Elms, A. C. (2009). Obedience lite. *American Psychologist, 64*(1), 32–36. https://doi.org/10.1037/a0014473

Elsner, B., & Pauen, S. (2012). Vorgeburtliche Entwicklung und früheste Kindheit (0-2 Jahre) [Prenatal development and earliest childhood (0-2 years)]. In W. Schneider & U. Lindenberger (Eds.), *Entwicklungspsychologie [developmental psychology]* (7th ed., pp. 159–185). Weinheim, Basel: Beltz.

Embgen, S., Luber, M., Becker-Asano, C., Ragni, M., Evers, V., & Arras, K. O. (2012). Robot-specific social cues in emotional body language. In I. Staff (Ed.), *2012 IEEE Ro-Man* (pp. 1019–1025). Piscataway, N.J: IEEE. https://doi.org/10.1109/ROMAN.2012.6343883

Endo, N., Momoki, S., Zecca, M., Saito, M., Mizoguchi, Y., Itoh, K., & Takanishi, A. (2008). Development of whole-body emotion expression humanoid robot. In *IEEE International Conference on Robotics and Automation, ICRA* (pp. 2140–2145). Piscataway, NJ: IEEE. https://doi.org/10.1109/ROBOT.2008.4543523

Eyssel, F. (2017). An experimental psychological perspective on social robotics. *Robotics and Autonomous Systems, 87*, 363–371. https://doi.org/10.1016/j.robot.2016.08.029

Faceplanted. (2014). Soldiers are developing relationships with their battlefield robots, naming them, assigining genders, and even holding funerals when they are destroyed [Online forum comment]. Retrieved from https://www.reddit.com/r/technology/comments/1mn6wo/soldiers_are_developing_relationships_with_their/

Fan, X., Miller, B. C., Park, K.-E., Winward, B. W., Christensen, M., Grotevant, H. D., & Tai, R. H. (2006). An Exploratory Study about Inaccuracy and Invalidity in Adolescent Self-Report Surveys. *Field Methods, 18*(3), 223–244. https://doi.org/10.1177/152822x06289161

Fasel, B., & Luettin, J. (2003). Automatic facial expression analysis: a survey. *Pattern Recognition, 36*(1), 259–275. https://doi.org/10.1016/S0031-3203(02)00052-3

Feil-Seifer, D., & Matarić, M. J. (2009). Human-Robot Interaction. In R. A. Meyers (Ed.), *Encyclopedia of complexity and systems science* (pp. 4643–4659). New York: Springer. https://doi.org/10.1007/978-0-387-30440-3_ 274

Field, A. (2013). *Discovering statistics using IBM SPSS statistics* (4th edition). London: Sage.

Field, A., Miles, J., & Field, Z. (2012). *Discovering statistics using R.* London: Sage.

Fiske, S. T., & Taylor, S. E. (1984). *Social cognition. Topics in social psychology.* New York, NY: Random House.

Fong, T., Nourbakhsh, I., & Dautenhahn, K. (2003). A survey of socially interactive robots. *Robotics and Autonomous Systems, 42*(3–4), 143–166. https://doi.org/10.1016/S0921-8890(02)00372-X

Fong, T., Thorpe, C., & Baur, C. (2003). Collaboration, Dialogue, Human-Robot Interaction. In R. A. Jarvis & A. Zelinsky (Eds.), *Springer Tracts in Advanced Robotics: Vol. 6. Robotics Research: The Tenth International Symposium* (Vol. 6, pp. 255–266). Berlin, Heidelberg: Springer. https://doi.org/10.1007/3-540-36460-9_17

Fontaine, J. R. J., Scherer, K. R., Roesch, E. B., & Ellsworth, P. C. (2007). The world of emotions is not two-dimensional. *Psychological Science, 18*(12), 1050–1057. https://doi.org/10.1111/j.1467-9280.2007.02024.x

Frank, M. G., Ekman, P., & Friesen, W. V. (1993). Behavioral markers and recognizability of the smile of enjoyment. *Journal of Personality and Social Psychology, 64*(1), 83–93.

Fridlund, A. J. (1991). Sociality of solitary smiling: Potentiation by an implicit audience. *Journal of Personality and Social Psychology, 60*(2), 229–240. https://doi.org/10.1037/0022-3514.60.2.229

Fridlund, A. J. (1992). The behavioral ecology and sociality of human faces. In M. S. Clark (Ed.), *Review of personality and social psychology, No. 13. Emotion* (pp. 90–121). Thousand Oaks, CA, US: Sage Publications, Inc.

Frijda, N. H. (2007). *The laws of emotion.* Mahwah, NJ: Erlbaum. Retrieved from http://www.loc.gov/catdir/enhancements/fy0622/2006298893-b.html

Frith, U., & Frith, C. D. (2003). Development and neurophysiology of mentalizing. *Philosophical Transactions of the Royal Society B: Biological Sciences, 358*(1431), 459–473. https://doi.org/10.1098/rstb.2002.1218

Fujita, M. (2004). On activating human communications with pet-type robot AIBO. *Proceedings of the IEEE, 92*(11), 1804–1813. https://doi.org/10.1109/JPROC.2004.835364

Garreau, J. (2007, May 6). Bots on The Ground. In the Field of Battle (Or Even Above It), Robots Are a Soldier's Best Friend. *The Washington Post.*

Geiskkovitch, D. Y., Cormier, D., Seo, S. H., & Young, J. E. (2016). Please Continue, We Need More Data: An Exploration of Obedience to Robots. *Journal of Human-Robot Interaction, 5*(1), 82. https://doi.org/10.5898/10.5898/JHRI.5.1.Geiskkovitch

Gelder, B. de. (2009). Why bodies? Twelve reasons for including bodily expressions in affective neuroscience. *Philosophical Transactions of the Royal Society B: Biological Sciences, 364*(1535), 3475–3484. https://doi.org/10.1098/rstb.2009.0190

Gerdes, A. B. M., Wieser, M. J., & Alpers, G. W. (2014). Emotional pictures and sounds: a review of multimodal interactions of emotion cues in multiple domains. *Frontiers in Psychology, 5*, 1351. https://doi.org/10.3389/fpsyg.2014.01351

Gergen, K. J. (1973). Social psychology as history. *Journal of Personality and Social Psychology, 26*(2), 309–320.

German Scientific Board. (2018). Pressemitteilungen: Psychologie in der Verantwortung | Wissenschaftsrat empfiehlt Öffnung der Psychologie und neue Wege in der Pychotherapieausbildung [Press releases. Psychology in responsibility | Science Council recommends opening psychology and new ways in pychotherapy education]. Retrieved from https://www.wissenschaftsrat.de/index.php?id=1414&print=1&L=

Germine, L., Nakayama, K., Duchaine, B. C., Chabris, C. F., Chatterjee, G., & Wilmer, J. B. (2012). Is the Web as good as the lab? Comparable performance from Web and lab in cognitive/perceptual experiments. *Psychonomic Bulletin & Review, 19*(5), 847–857. https://doi.org/10.3758/s13423-012-0296-9

Gonsior, B., Sosnowski, S., Buss, M., Wollherr, D., & Kuhnlenz, K. (2012). An emotional adaption approach to increase helpfulness towards a robot. In *IEEE/RSJ International Conference on Intelligent Robots and Systems (IROS)* (pp. 2429–2436). Piscataway, NJ: IEEE. https://doi.org/10.1109/IROS.2012.6385941

Gonsior, B., Sosnowski, S., Mayer, C., Blume, J., Radig, B., Wollherr, D., & Kuhnlenz, K. (2011). Improving aspects of empathy and subjective performance for HRI through mirroring facial expressions. In M. Scheutz (Ed.), *IEEE RO-MAN, 2011: International Symposium on Robot and Human Interactive Communication* (pp. 350–356). Piscataway, NJ: IEEE. https://doi.org/10.1109/ROMAN.2011.6005294

Greenwood, J. D. (1982). On the Relation Between Laboratory Experiments and Social Behaviour: Causal Explanation and Generalization. *Journal for the Theory of Social Behaviour, 12*(3), 225–250. https://doi.org/10.1111/j.1468-5914.1982.tb00449.x

Griskevicius, V., Shiota, M. N., & Neufeld, S. L. (2010). Influence of different positive emotions on persuasion processing: a functional evolutionary approach. *Emotion, 10*(2), 190–206. https://doi.org/10.1037/a0018421

Gross, J. J. (1998). The emerging field of emotion regulation: An integrative review. *Review of General Psychology, 2*(3), 271–299. https://doi.org/10.1037/1089-2680.2.3.271

Gross, J. J. (2010). The Future's So Bright, I Gotta Wear Shades. *Emotion Review, 2*(3), 212–216. https://doi.org/10.1177/1754073910361982

Gross, J. J., & Levenson, R. W. (1995). Emotion elicitation using films. *Cognition & Emotion, 9*(1), 87–108. https://doi.org/10.1080/02699939508408966

Haidt, J., & Keltner, D. (1999). Culture and Facial Expression: Open-ended Methods Find More Expressions and a Gradient of Recognition. *Cognition & Emotion, 13*(3), 225–266. https://doi.org/10.1080/026999399379267

Ham, J., & Midden, C. J. H. (2014). A Persuasive Robot to Stimulate Energy Conservation: The Influence of Positive and Negative Social Feedback and Task Similarity on Energy-Consumption Behavior. *International Journal of Social Robotics*, 6(2), 163–171. https://doi.org/10.1007/s12369-013-0205-z

Häring, M., Bee, N., & André, E. (2011). Creation and Evaluation of emotion expression with body movement, sound and eye color for humanoid robots. In M. Scheutz (Ed.), *IEEE RO-MAN, 2011: International Symposium on Robot and Human Interactive Communication* (pp. 204–209). Piscataway, NJ: IEEE. https://doi.org/10.1109/ROMAN.2011.6005263

Harris, C., & Alvarado, N. (2005). Facial expressions, smile types, and self-report during humour, tickle, and pain. *Cognition & Emotion*, 19(5), 655–669. https://doi.org/10.1080/02699930441000472

Haslam, S. A., & Reicher, S. D. (2017). 50 Years of "Obedience to Authority": From Blind Conformity to Engaged Followership. *Annual Review of Law and Social Science*, 13(1), 59–78. https://doi.org/10.1146/annurev-lawsocsci-110316-113710

Haslam, S. A., Reicher, S. D., & Millard, K. (2015). Shock Treatment: Using Immersive Digital Realism to Restage and Re-examine Milgram's 'Obedience to Authority' Research. *PloS One*, 10(3). https://doi.org/10.1371/journal.pone.0109015

Hatfield, E., Cacioppo, J. T., & Rapson, R. L. (1992). Primitive emotional contagion. In M. S. Clark (Ed.), *Review of personality and social psychology, No. 13. Emotion* (pp. 151–177). Thousand Oaks, CA, US: Sage Publications, Inc.

Haxby, J. V., Hoffman, E. A., & Gobbini, M.I. (2000). The distributed human neural system for face perception. *Trends in Cognitive Sciences*, 4(6), 223–233. https://doi.org/10.1016/s1364-6613(00)01482-0

Hayes, B., Ullman, D., Alexander, E., Bank, C., & Scassellati, B. (2014). People help robots who help others, not robots who help themselves. In *The 23rd IEEE International Symposium on Robot and Human Interactive Communication* (pp. 255–260). IEEE. https://doi.org/10.1109/ROMAN.2014.6926262

Heerink, M., Kröse, B., Evers, V., & Wielinga, B. (2010). Assessing Acceptance of Assistive Social Agent Technology by Older Adults: the Almere Model. *International Journal of Social Robotics*, *2*(4), 361–375. https://doi.org/10. 1007/s12369-010-0068-5

Heerink, M., Kröse, B., Evers, V., & Wielinga, B. (2010). Relating conversational expressiveness to social presence and acceptance of an assistive social robot. *Virtual Reality*, *14*(1), 77–84. https://doi.org/10.1007/s10055-009-0142-1

Hegel, F., Eyssel, F., & Wrede, B. (2010). The social robot 'Flobi': Key concepts of industrial design. In *Proceedings of the 19th International Symposium in Robot and Human Interactive Communication (RO-MAN 2010)* (pp. 120–125). IEEE. https://doi.org/10.1109/ROMAN.2010.5598691

Hegel, F., Spexard, T., Wrede, B., Horstmann, G., & Vogt, T. (2006). Playing a different imitation game: Interaction with an Empathic Android Robot. In *6th IEEE-RAS International Conference on Humanoid Robots, 2006* (pp. 56–61). Piscataway, NJ: IEEE Operations Center. https://doi.org/10. 1109/ICHR.2006.321363

Heider, F., & Simmel, M. (1944). An experimental study of apparent behavior. *The American Journal of Psychology*, *57*(2), 243–259.

Hendrick, C. (1990). Replications, strict replications, and conceptual replications: Are they important? *Journal of Social Behavior & Personality*, *5*(4), 41–49.

Hess, U., Banse, R., & Kappas, A. (1995). The intensity of facial expression is determined by underlying affective state and social situation. *Journal of Personality and Social Psychology*, *69*(2), 280–288. https://doi.org/10.1037/ 0022-3514.69.2.280

Hinton, P. R., Brownlow, C., McMurray, I., & Cozens, B. (2004). *SPSS Explained* (1st ed.). London, New York: Routledge.

Hoffman, M. L. (1975). Developmental synthesis of affect and cognition and its implications for altruistic motivation. *Developmental Psychology*, *11*(5), 607–622. https://doi.org/10.1037/0012-1649.11.5.607

Hoffman, M. L. (1982). Development of prosocial motivation: Empathy and guilt. In N. Eisenberg (Ed.), *The development of prosocial behavior* (pp. 281–313). New York: Academic Press.

Hoffman, M. L. (2008). Empathy and prosocial behavior. In M. Lewis, J. M. Haviland-Jones, & L. Feldman Barrett (Eds.), *Handbook of emotions* (3rd ed., pp. 440–455). New York: Guilford Press.

Hooge, I. E. de, Zeelenberg, M., & Breugelmans, S. M. (2007). Moral sentiments and cooperation: Differential influences of shame and guilt. *Cognition & Emotion*, *21*(5), 1025–1042. https://doi.org/10.1080/02699930600980874

Horstmann, A. C., Bock, N., Linhuber, E., Szczuka, J. M., Straßmann, C., & Krämer, N. C. (2018). Do a robot's social skills and its objection discourage interactants from switching the robot off? *PloS One*, *13*(7), e0201581. https://doi.org/10.1371/journal.pone.0201581

Hsee, C. K., Hatfield, E., Carlson, J. G., & Chemtob, C. (1990). The effect of power on susceptibility to emotional contagion. *Cognition & Emotion*, *4*(4), 327–340. https://doi.org/10.1080/02699939008408081

Hughes, C., & Dunn, J. (2002). 'When I say a naughty word'. A longitudinal study of young children's accounts of anger and sadness in themselves and close others. *British Journal of Developmental Psychology*, *20*(4), 515–535. https://doi.org/10.1348/026151002760390837

Hussy, W., & Jain, A. (2002). *Experimentelle Hypothesenprüfung in der Psychologie [Experimental hypotheses testing in psychology]*. Göttingen: Hogrefe.

InmemoryofRomeo. (2011b). Pleakley learns to sing [Video Clip on YouTube]. Retrieved from https://www.youtube.com/watch?v=O0xWOtO4mBk

InmemoryofRomeo. (2011a). Pleo RB Learning Stone Demonstration [Video Clip on YouTube]. Retrieved from https://www.youtube.com/watch?v=wnsnNP-A6j8

Innvo Labs. (2012). Pleo rb. Retrieved from https://www.pleoworld.com/pleo_rb/eng/index.php

IRobot. (2019). Roomba Robot Vacuums. Retrieved from https://www.irobot.ie/Home-Robots/Vacuuming

Izard, C. E. (1992). Basic emotions, relations among emotions, and emotion-cognition relations. *Psychological Review*, *99*(3), 561–565.

Izard, C. E., Dougherty, F. E., Bloxom, B. M., & Kotsch, N. E. (1974). *The Differential Emotions Scale: a method of measuring the subjective experience of discrete emotions.* Nashville: Vanderbilt University.

Izard, C. E. (2010). The Many Meanings/Aspects of Emotion: Definitions, Functions, Activation, and Regulation. *Emotion Review, 2*(4), 363–370. https://doi.org/10.1177/1754073910374661

Jaeckel, P., Campbell, N., & Melhuish, C. (2008). Facial behaviour mapping-From video footage to a robot head. *Robotics and Autonomous Systems, 56*(12), 1042–1049. https://doi.org/10.1016/j.robot.2008.09.002

Jakobs, E., Manstead, A. S. R., & Fischer, A. H. (2001). Social context effects on facial activity in a negative emotional setting. *Emotion, 1*(1), 51–69. https://doi.org/10.1037/1528-3542.1.1.51

James, W. (1884). What is an emotion? *Mind, 9*(34), 188–205.

Jiang, B., Martinez, B., & Pantic, M. (2014). Parametric temporal alignment for the detection of facial action temporal segments. In M. Valstar, A. French, & T. Pridmore (Eds.), *Proceedings of the British Machine Vision Conference 2014* (102.1-102.11). British Machine Vision Association. https://doi.org/10.5244/C.28.102

Jo, D., Han, J., Chung, K., & Lee, S. (2013). Empathy between human and robot? In *Proceedings of the 8th ACM/IEEE International Conference on Human-Robot Interaction* (pp. 151–152). Piscataway, NJ: IEEE. https://doi.org/10.1109/HRI.2013.6483546

Johansson, G., Hofsten, C. v. H., & Jansson, G. (1980). Event perception. *Annual Review of Psychology, 31*(1), 27–63.

Kahn, P. H. (1997). Developmental Psychology and the Biophilia Hypothesis: Children's Affiliation with Nature. *Developmental Review, 17*(1), 1–61. https://doi.org/10.1006/drev.1996.0430

Kahn, P. H., Friedman, B., Pérez-Granados, D. R., & Freier, N. G. (2006). Robotic pets in the lives of preschool children. *Interaction Studies, 7*(3), 405–436. https://doi.org/10.1075/is.7.3.13kah

Kahn, P. H., Kanda, T., Ishiguro, H., Freier, N. G., Severson, R. L., Gill, B. T., . . . Shen, S. (2012). "Robovie, you'll have to go into the closet now": children's social and moral relationships with a humanoid robot. *Developmental Psychology, 48*(2), 303–314. https://doi.org/10.1037/a0027033

Kaltwang, S., Todorovic, S., & Pantic, M. (2015). Latent trees for estimating intensity of facial action units. *Computer Vision and Pattern Recognition,* 296–304.

Kanwisher, N., McDermott, J., & Chun, M. M. (1997). The fusiform face area: a module in human extrastriate cortex specialized for face perception. *The Journal of Neuroscience : the Official Journal of the Society for Neuroscience, 17*(11), 4302–4311.

Kappas, A., Hess, U., & Scherer, K. R. (1991). Voice and emotion. In R. S. Feldman & B. Rimé (Eds.), *Studies in emotion & social interaction. Fundamentals of nonverbal behavior* (pp. 200–238). New York: Cambridge University Press.

Kappas, A., Krumhuber, E., Küster, D. (2013). Facial behavior. In J. A. Hall & M. L. Knapp (Eds.), *Handbooks of Communication Sciences: Nonverbal Communication* (pp. 131–165). Berlin: De Gruyter.

Kawulok, M., Celebi, M. E., & Smołka, B. (Eds.). (2016). *Advances in face detection and facial image analysis.* Cham: Springer.

Kelly, J. R., & Barsade, S. G. (2001). Mood and Emotions in Small Groups and Work Teams. *Organizational Behavior and Human Decision Processes, 86*(1), 99–130. https://doi.org/10.1006/obhd.2001.2974

Keltner, D., & Bonanno, G. A. (1997). A study of laughter and dissociation: distinct correlates of laughter and smiling during bereavement. *Journal of Personality and Social Psychology, 73*(4), 687–702.

Keltner, D., Ellsworth, P. C., & Edwards, K. (1993). Beyond simple pessimism: effects of sadness and anger on social perception. *Journal of Personality and Social Psychology, 64*(5), 740–752.

Keppel, G. (1991). *Design and analysis. A researcher's handbook.* Englewood Cliffs, N.J.: Pearson Prentice Hall.

Kim, E. H., Kwak, S. S., Hyun, K. H., Kim, S. H., & Kwak, Y. K. (2009). Design and Development of an Emotional Interaction Robot, Mung. *Advanced Robotics*, *23*(6), 767–784. https://doi.org/10.1163/156855309X431712

Kim, T., & Hinds, P. (2006). Who Should I Blame? Effects of Autonomy and Transparency on Attributions in Human-Robot Interaction. In *The 15th IEEE International Symposium on Robot and Human Interactive Communication, RO-MAN* (pp. 80–85). Piscataway, NJ: IEEE. https://doi.org/10. 1109/ROMAN.2006.314398

Kirby, R., Forlizzi, J., & Simmons, R. (2010). Affective social robots. *Robotics and Autonomous Systems*, *58*(3), 322–332. https://doi.org/10.1016/j.robot. 2009.09.015

Kleinginna, P. R., & Kleinginna, A. M. (1981). A categorized list of emotion definitions, with suggestions for a consensual definition. *Motivation and Emotion*, *5*(4), 345–379. https://doi.org/10.1007/BF00992553

Ko, B. C. (2018). A Brief Review of Facial Emotion Recognition Based on Visual Information. *Sensors*, *18*(2), pii: E401. https://doi.org/10.3390/s18020401

Kozima, H., Michalowski, M. P., & Nakagawa, C. (2009). Keepon. *International Journal of Social Robotics*, *1*(1), 3–18. https://doi.org/10.1007/s12369-008-0009-8

Krach, S., Hegel, F., Wrede, B., Sagerer, G., Binkofski, F., & Kircher, T. (2008). Can machines think? Interaction and perspective taking with robots investigated via fMRI. *PloS One*, *3*(7), e2597. https://doi.org/10.1371/journal.pone.0002597.

Krämer, N., Klatt, J., Hoffmann, L., & Rosenthal-von der Pütten, A. M. (2013). "Emotional" robots and agents-implementation of emotions in artificial entities. In C. Mohiyeddini, M. W. Eysenck, & S. Bauer (Eds.), *Psychology of emotions, motivations and actions. Handbook of psychology of emotions: Recent theoretical perspectives and novel empirical findings* (2nd ed., pp. 277–296). New York: Nova publishers.

Krämer, N. C., Eimler, S. C., von der Pütten, A. M., & Payr, S. (2011). Theory of companions: what can theoretical models contribute to applications and understanding of human-robot interaction? *Applied Artificial Intelligence*, *25*(6), 474–502.

Krämer, N. C., Lucas, G., Schmitt, L., & Gratch, J. (2018). Social snacking with a virtual agent – On the interrelation of need to belong and effects of social responsiveness when interacting with artificial entities. *International Journal of Human-Computer Studies, 109*(C), 112–121. https://doi.org/10. 1016/j.ijhcs.2017.09.001

Krebs, D. (1975). Empathy and altruism. *Journal of Personality and Social Psychology, 32*(6), 1134–1146. https://doi.org/10.1037/0022-3514.32.6.1134

Kring, A. M., & Gordon, A. H. (1998). Sex differences in emotion: expression, experience, and physiology. *Journal of Personality and Social Psychology, 74*(3), 686–703.

Kring, A. M., & Sloan, D. M. (2007). The Facial Expression Coding System (FACES): development, validation, and utility. *Psychological Assessment, 19*(2), 210–224. https://doi.org/10.1037/1040-3590.19.2.210

Kringelbach, M. L., Stark, E. A., Alexander, C., Bornstein, M. H., & Stein, A. (2016). On Cuteness: Unlocking the Parental Brain and Beyond. *Trends in Cognitive Sciences, 20*(7), 545–558. https://doi.org/10.1016/j.tics.2016. 05.003

Kubinger, K. D., Rasch, D., & Moder, K. (2009). Zur Legende der Voraussetzungen des t-Tests für unabhängige Stichproben [About the legend of the requirements for the t-test for independent samples]. *Psychologische Rundschau, 60*(1), 26–27. https://doi.org/10.1026/0033-3042.60.1. 26

Kulic, D., & Croft, E. (2007). Physiological and subjective responses to articulated robot motion. *Robotica, 25*(01), 13. https://doi.org/10.1017/ S0263574706002955

Kwak, S. S., Kim, Y., Kim, E., Shin, C., & Cho, K. (2013). What makes people empathize with an emotional robot?: The impact of agency and physical embodiment on human empathy for a robot. In *The 22nd IEEE International Symposium on Robot and Human Interactive Communication, IEEE RO-MAN* (pp. 180–185). Piscataway, NJ: IEEE. https://doi.org/10.1109/ ROMAN.2013.6628441

Lamm, H., & Stephan, E. (1986). Zur Messung von Einsamkeit: Entwicklung einer deutschen Fassung des Fragebogens von Russell und Peplau [Measuring Loneliness: Development of a german version of the Russell and Peplau questionnaire]. *Zeitschrift Für Arbeits- Und Organisationspsychologie, 30*, 132–134.

Lane, R. D., Ahern, G. L., Schwartz, G. E., & Kaszniak, A. W. (1997). Is Alexithymia the Emotional Equivalent of Blindsight? *Biological Psychiatry, 42*(9), 834–844. https://doi.org/10.1016/S0006-3223(97)00050-4

Lang, P. J., Greenwald, M. K., Bradley, M. M., & Hamm, A. O. (1993). Looking at pictures: affective, facial, visceral, and behavioral reactions. *Psychophysiology, 30*(3), 261–273.

Lang, P. J., Bradley, M. M., & Cuthbert, B. N. (1997). Motivated attention: Affect, activation, and action. In P. J. Lang, R. F. Simons, & M. T. Balaban (Eds.), *Attention and orienting: Sensory and motivational processes* (pp. 97–135). Mahwah, NJ, US: Lawrence Erlbaum Associates Publishers.

Lapakko, D. (1997). Three cheers for language: A closer examination of a widely cited study of nonverbal communication. *Communication Education, 46*(1), 63–67. https://doi.org/10.1080/03634529709379073

Larsen, J. T., McGraw, A. P., & Cacioppo, J. T. (2001). Can people feel happy and sad at the same time? *Journal of Personality and Social Psychology, 81*(4), 684–696.

Larsen, J. T., Norris, C. J., & Cacioppo, J. T. (2003). Effects of positive and negative affect on electromyographic activity over zygomaticus major and corrugator supercilii. *Psychophysiology, 40*(5), 776–785. https://doi.org/10.1111/1469-8986.00078

Larsen, J.T., Berntson, G. G., Poehlmann, K. M., Ito, T. A., & Cacioppo, J.T. (2008). The psychophysiology of emotion. In M. Lewis, J. M. Haviland-Jones, & L. Feldman Barrett (Eds.), *Handbook of emotions* (3rd ed.). New York: Guilford Press.

Larzelere, R. E., Kuhn, B. R., & Johnson, B. (2004). The intervention selection bias: an underrecognized confound in intervention research. *Psychological Bulletin, 130*(2), 289–303. https://doi.org/10.1037/0033-2909.130.2.289

Lazarus, R. S. (1991). Progress on a cognitive-motivational-relational theory of emotion. *American Psychologist, 46*(8), 819–834. https://doi.org/10.1037/0003-066x.46.8.819

Lee, H. S., Park, J. W., Jo, S. H., & Chung, M. J. (2007). A Linear Dynamic Affect-Expression Model: Facial Expressions According to Perceived Emotions in Mascot-Type Facial Robots. In *16th IEEE International Conference on Robot & Human Interactive Communication* (pp. 619–624). Piscataway, N.J.: IEEE. https://doi.org/10.1109/ROMAN.2007.4415158

Lee, V., & Wagner, H. (2002). The effect of social presence on the facial and verbal expression of emotion and the interrelationships among emotion components. *Journal of Nonverbal Behavior, 26*(1), 3–25. https://doi.org/10.1023/A:1014479919684

Leite, I., Castellano, G., Pereira, A., Martinho, C., & Paiva, A. (2014). Empathic Robots for Long-term Interaction. *International Journal of Social Robotics, 6*(3), 329–341. https://doi.org/10.1007/s12369-014-0227-1

Leite, I., Martinho, C., & Paiva, A. (2013). Social Robots for Long-Term Interaction: A Survey. *International Journal of Social Robotics, 5*(2), 291–308. https://doi.org/10.1007/s12369-013-0178-y

Leite, I., Martinho, C., Pereira, A., & Paiva, A. (2008). iCat: an affective game buddy based on anticipatory mechanisms. In *Proceedings of the 7th international joint conference on Autonomous agents and multiagent systems - Volume 3* (pp. 1229–1232). Estoril, Portugal: International Foundation for Autonomous Agents and Multiagent Systems.

Leite, I., Mascarenhas, S., Pereira, A., Martinho, C., Prada, R., & Paiva, A. (2010). "Why Can't We Be Friends?" An Empathic Game Companion for Long-Term Interaction. In J. Allbeck, N. Badler, T. Bickmore, C. Pelachaud, & A. Safonova (Eds.), *Lecture notes in computer science Lecture notes in artificial intelligence: Vol. 6356. Intelligent virtual agents: 10th international conference, IVA 2010, Philadelphia, PA, USA, September 20 - 22, 2010 ; proceedings* (Vol. 6356, pp. 315–321). Berlin: Springer. https://doi.org/10.1007/978-3-642-15892-6_32

Lesser, I. M. (1981). A Review of the Alexithymia Concept. *Psychosomatic Medicine, 43*(6), 531–543. https://doi.org/10.1097/00006842-198112000-00009

Levenson, R. W. (1996). Biological substrates of empathy and facial modulation of emotion: Two facets of the scientific legacy of John Lanzetta. *Motivation and Emotion*, *20*(3), 185–204. https://doi.org/10.1007/BF02251886

Levenson, R. W., & Ruef, A. M. (1992). Empathy: A physiological substrate. *Journal of Personality and Social Psychology*, *63*(2), 234–246. https://doi.org/10.1037/0022-3514.63.2.234

Levy, G. S., Angel-levy, P., Levy, E. J., Levy, S. A., & Levy, J. A. 7949616.

Lewis, I., Watson, B., & White, K. M. (2009). Internet versus paper-and-pencil survey methods in psychological experiments: Equivalence testing of participant responses to health-related messages. *Australian Journal of Psychology*, *61*(2), 107–116. https://doi.org/10.1080/00049530802105865

Lewis, M., Haviland-Jones, J. M., & Feldman Barrett, L. (Eds.). (2008). *Handbook of emotions* (3. ed.). New York: Guilford Press.

Li, J., & Chignell, M. (2011). Communication of Emotion in Social Robots through Simple Head and Arm Movements. *International Journal of Social Robotics*, *3*(2), 125–142. https://doi.org/10.1007/s12369-010-0071-x

Lien, J.-J. J., Kanade, T., Cohn, J., & Li, C. (2000). Detection, tracking, and classification of subtle changes in facial expression. *Journal of Robotics and Autonomous Systems*, *31*, 131–146.

Lin, P., Abney, K., & Bekey, G. A. (Eds.). (2014). *Intelligent robotics and autonomous agents. Robot ethics: The ethical and social implications of robotics* (First MIT Press paperback edition). Cambridge, Massachusetts, London, England: The MIT Press.

Lindquist, K. A. (2013). Emotions Emerge from More Basic Psychological Ingredients: A Modern Psychological Constructionist Model. *Emotion Review*, *5*(4), 356–368. https://doi.org/10.1177/1754073913489750

Lisetti, C., & Hudlicka, E. (2015). Why and how to build emotion-based agent architectures. In R. A. Calvo, S. D'Mello, J. Gratch, & A. Kappas (Eds.), *Oxford library of psychology. The Oxford handbook of affective computing* (pp. 94–109). Oxford, New York: Oxford University Press.

Littlewort, G., Bartlett, M.S., Fasel, I.R., Chenu, J., Kanda, T., Ishiguro, H., & Movellan, J.R. (2004). Towards Social Robots: Automatic Evaluation of Human-robot Interaction by Face Detection and Expression Classification - Semantic Scholar. *NIPS*.

Liu, S., Helfenstein, S., & Wahlstedt, A. (2008). Social psychology of persuasion applied to a human agent interaction. *Human Technology*, 123–143. Retrieved from https://jyx.jyu.fi/dspace/handle/123456789/20224

López, G., Quesada, L., & Guerrero, L. A. (2018). Alexa vs. Siri vs. Cortana vs. Google Assistant: A Comparison of Speech-Based Natural User Interfaces. In I. L. Nunes (Ed.), *Advances in Intelligent Systems and Computing: Vol. 592. Advances in Human Factors and Systems Interaction: Proceedings of the AHFE 2017 International Conference on Human Factors and Systems Interaction* (Vol. 592, pp. 241–250). Cham: Springer International Publishing. https://doi.org/10.1007/978-3-319-60366-7_23

Lorenz, K. (1943). Die angeborenen Formen möglichen Verhaltens [The innate forms of possible behavior]. *Zeitschrift Für Tierpsychologie, 5*, 235–409.

Maner, J. K., Kenrick, D. T., Becker, D. V., Robertson, T. E., Hofer, B., Neuberg, S. L., . . . Schaller, M. (2005). Functional projection: how fundamental social motives can bias interpersonal perception. *Journal of Personality and Social Psychology, 88*(1), 63–78. https://doi.org/10.1037/0022-3514.88.1.63

Marsella, S., Gratch, J., & Petta, P. (2010). Computational models of emotion. In K. R. Scherer, T. Bänziger, & E. B. Roesch (Eds.), *Series in affective science. Blueprint for affective computing: A sourcebook* (pp. 21–46). Oxford: Oxford Univ. Press.

Martin, J., Rychlowska, M., Wood, A., & Niedenthal, P. (2017). Smiles as Multipurpose Social Signals. *Trends in Cognitive Sciences, 21*(11), 864–877. https://doi.org/10.1016/j.tics.2017.08.007

Martinez, B., & Valstar, M. F. (2016). Advances, Challenges, and Opportunities in Automatic Facial Expression Recognition. In M. Kawulok, M. E. Celebi, & B. Smołka (Eds.), *Advances in face detection and facial image analysis* (pp. 63–100). Cham: Springer. https://doi.org/10.1007/978-3-319-25958-1_4

Matarić, M., Tapus, A., Winstein, C., & Eriksson, J. (2009). Socially assistive robotics for stroke and mild TBI rehabilitation. *Studies in Health Technology and Informatics*, *145*, 249–262.

Matsumoto, D., Keltner, D., Shiota, M. N., O'Sullivan, M., & Frank, M. (2008). Facial Expressions of Emotion. In M. Lewis, J. M. Haviland-Jones, & L. Feldman Barrett (Eds.), *Handbook of emotions* (3rd ed., pp. 211–234). New York: Guilford Press.

Mauss, I. B., Levenson, R. W., McCarter, L., Wilhelm, F. H., & Gross, J. J. (2005). The tie that binds? Coherence among emotion experience, behavior, and physiology. *Emotion (Washington, D.C.)*, *5*(2), 175–190. https://doi.org/10.1037/1528-3542.5.2.175

Mauss, I. B., & Robinson, M. D. (2009). Measures of emotion: A review. *Cognition & Emotion*, *23*(2), 209–237. https://doi.org/10.1080/02699930802204677

Mehrabian, A. (1976). Questionnaire measures of affiliative tendency and sensitivity to rejection. *Psychological Reports*, *38*(1), 199–209.

Mehrabian, A. (1970). The Development and Validation of Measures of Affiliative Tendency and Sensitivity To Rejection. *Educational and Psychological Measurement*, *30*(2), 417–428. https://doi.org/10.1177/001316447003000226

Mehrabian, A. (1981). *Silent messages: Implicit communication of emotions and attitudes* (2d ed.). Belmont, Calif.: Wadsworth Pub. Co.

Mele, A. (1995). *Autonomous Agents*. New York: Oxford University Press.

Melson, G. F., Kahn, J. P. H., Beck, A., & Friedman, B. (2009). Robotic Pets in Human Lives: Implications for the Human-Animal Bond and for Human Relationships with Personified Technologies. *Journal of Social Issues*, *65*(3), 545–567. https://doi.org/10.1111/j.1540-4560.2009.01613.x

Menne, I. M. (2017). Yes, of Course? An Investigation on Obedience and Feelings of Shame Towards a Robot. In A. Kheddar, E. Yoshida, S. S. Ge, K. Suzuki, J.-J. Cabibihan, F. Eyssel, & H. He (Eds.), *Lecture Notes in Computer Science: Vol. 10652. Social Robotics: 9th International Conference, ICSR 2017* (Vol. 10652, pp. 365–374). Cham: Springer International Publishing. https://doi.org/10.1007/978-3-319-70022-9_36

Menne, I. M., & Lugrin, B. (2017). In the Face of Emotion: A Behavioral Study on Emotions Towards a Robot Using the Facial Action Coding System. In *Companion of the 2017 ACM/IEEE International Conference on Human-Robot Interaction, HRI 2017* (pp. 205–206). https://doi.org/10.1145/3029798.3038375

Menne, I. M., Schnellbacher, C., & Schwab, F. (2016). Facing Emotional Reactions Towards a Robot – An Experimental Study Using FACS. In A. Agah, J.-J. Cabibihan, A. M. Howard, M. A. Salichs, & H. He (Eds.), *Lecture Notes in Computer Science. Social Robotics* (Vol. 9979, pp. 372–381). Cham: Springer International Publishing. https://doi.org/10.1007/978-3-319-47437-3_36

Menne, I. M., & Schwab, F. (2018). Faces of Emotion: Investigating Emotional Facial Expressions Towards a Robot. *International Journal of Social Robotics, 10*(2), 199–209. https://doi.org/10.1007/s12369-017-0447-2

Merten, J. (2003). *Einführung in die Emotionspsychologie [introduction to emotion psychology]* (1. Aufl.). Stuttgart: Kohlhammer.

Mesquita, B., & Boiger, M. (2014). Emotions in Context: A Sociodynamic Model of Emotions. *Emotion Review, 6*(4), 298–302. https://doi.org/10.1177/1754073914534480

Meuleman, B., & Scherer, K. (2013). Nonlinear Appraisal Modeling: An Application of Machine Learning to the Study of Emotion Production. *IEEE Transactions on Affective Computing, 4*(4), 398–411. https://doi.org/10.1109/T-AFFC.2013.25

Meyer, M. L., Masten, C. L., Ma, Y., Wang, C., Shi, Z., Eisenberger, N. I., & Han, S. (2013). Empathy for the social suffering of friends and strangers recruits distinct patterns of brain activation. *Social Cognitive and Affective Neuroscience, 8*(4), 446–454. https://doi.org/10.1093/scan/nss019

Milgram, S. (1965a). *Obedience* [film]. University Park, PA: Penn State Audio-Visual Services.

Milgram, S. (1974). *Obedience to Authority. An Experimental View.* New York: Harper & Row.

Milgram, S. (1965b). Some Conditions of Obedience and Disobedience to Authority. *Human Relations, 18*(1), 57–76. https://doi.org/10.1177/001872676501800105

Milgram, S. (1963). Behavioral study of obedience. *The Journal of Abnormal and Social Psychology*, *67*(4), 371.

Milgram, S. (1983). Reflections on Morelli's "Dilemma of Obedience". *Metaphilosophy*, *14*(3–4), 190–194. https://doi.org/10.1111/j.1467-9973. 1983.tb00308.x

Millard, K. (2015). *Shock Room [documentary]*. Sydney: Charlie Prod.

Millard, K. (2014). Revisioning Obedience: Exploring the Role of Milgram's Skills as a Filmmaker in Bringing His Shocking Narrative to Life. *Journal of Social Issues*, *70*(3), 439–455. https://doi.org/10.1111/josi.12070

Mirnig, N., Strasser, E., Weiss, A., Kühnlenz, B., Wollherr, D., & Tscheligi, M. (2015). Can You Read My Face? *International Journal of Social Robotics*, *7*(1), 63–76. https://doi.org/10.1007/s12369-014-0261-z

Mitchell, J. P., Banaji, M. R., & Macrae, C. N. (2005). The link between social cognition and self-referential thought in the medial prefrontal cortex. *Journal of Cognitive Neuroscience*, *17*(8), 1306–1315. https://doi.org/10. 1162/0898929055002418

Mitchell, T. R., Thompson, L., Peterson, E., & Cronk, R. (1997). Temporal Adjustments in the Evaluation of Events: The "Rosy View". *Journal of Experimental Social Psychology*, *33*(4), 421–448. https://doi.org/10.1006/jesp. 1997.1333

Moors, A. (2009). Theories of emotion causation: A review. *Cognition & Emotion*, *23*(4), 625–662. https://doi.org/10.1080/02699930802645739

Morelli, M. (1983). Milgram's dilemma of obedience. *Metaphilosophy*, *14*(3/4), 183–189. Retrieved from http://www.jstor.org/stable/24435438

Moshkina, L. (2012). Improving request compliance through robot affect. In *AAAI conference on artificial intelligence* (pp. 2031–2037).

Mutlu, B., Yamaoka, F., Kanda, T., Ishiguro, H., & Hagita, N. (2009). Nonverbal leakage in robots: communication of intentions through seemingly unintentional behavior. In *Proceedings of the 4th ACM/IEEE international conference on Human robot interaction* (pp. 69–76). La Jolla, California, USA: ACM. https://doi.org/10.1145/1514095.1514110

Nass, C., & Moon, Y. (2000). Machines and Mindlessness: Social Responses to Computers. *Journal of Social Issues, 56*(1), 81–103. https://doi.org/10. 1111/0022-4537.00153

Neumann, D. L., Chan, R. C.K., Boyle, G. J., Wang, Y., & Westbury, H. R. (2015). Measures of Empathy. In D. H. Saklofske, G. Matthews, & G. J. Boyle (Eds.), *Measures of personality and social psychological constructs* (pp. 257–289). Amsterdam: Elsevier Academic Press. https://doi.org/10. 1016/B978-0-12-386915-9.00010-3

Niedenthal, P. M., Krauth-Gruber, S., & Ric, F. (2006). *Psychology of emotion: Interpersonal, experiential, and cognitive approaches. Principles of social psychology.* New York: Psychology Press.

Nieding, G., & Ohler, P. (2018). Medien und Entwicklung [media and development]. In W. Schneider & U. Lindenberger (Eds.), *Entwicklungspsychologie [developmental psychology]* (8th ed., pp. 729–743). Weinheim, Basel: Beltz.

Nijssen, S. R. R., Müller, B. C. N., van Baaren, R. B., & Paulus, M. (2019). Saving the Robot or the Human? Robots Who Feel Deserve Moral Care. *Social Cognition, 37*(1), 41-S2. https://doi.org/10.1521/soco.2019.37.1.41

Nomura, T., Uratani, T., Kanda, T., Matsumoto, K., Kidokoro, H., Suehiro, Y., & Yamada, S. (2015). Why Do Children Abuse Robots? In J. A. Adams (Ed.), *Proceedings of the Tenth Annual ACMIEEE International Conference on Human-Robot Interaction* (pp. 63–64). New York, NY: ACM. https:// doi.org/10.1145/2701973.2701977

Öhman, A. (2002). Automaticity and the Amygdala: Nonconscious Responses to Emotional Faces. *Current Directions in Psychological Science, 11*(2), 62–66. https://doi.org/10.1111/1467-8721.00169

Olderbak, S., Hildebrandt, A., Pinkpank, T., Sommer, W., & Wilhelm, O. (2014). Psychometric challenges and proposed solutions when scoring facial emotion expression codes. *Behavior Research Methods, 46*(4), 992–1006. https://doi.org/10.3758/s13428-013-0421-3

Orne, M. T., & Holland, C. H. (1968). On the ecological validity of laboratory deceptions. *International Journal of Psychiatry, 6*, 282–293.

Ortony, A., Clore, G. L., & Collins, A. (1999). *The cognitive structure of emotions* (Reprinted.). Cambridge: Cambridge Univ. Press.

Paiva, A., Leite, I., Boukricha, H., & Wachsmuth, I. (2017). Empathy in Virtual Agents and Robots. *ACM Transactions on Interactive Intelligent Systems*, 7(3), 1–40. https://doi.org/10.1145/2912150

Pantic, M., & Rothkrantz, L.J.M. (2000). Expert system for automatic analysis of facial expressions. *Image and Vision Computing*, 18(11), 881–905. https://doi.org/10.1016/S0262-8856(00)00034-2

Paolacci, G., & Chandler, J. (2014). Inside the Turk. *Current Directions in Psychological Science*, 23(3), 184–188. https://doi.org/10.1177/0963721414531598

Papousek, I., Schulter, G., & Lang, B. (2009). Effects of emotionally contagious films on changes in hemisphere-specific cognitive performance. *Emotion*, 9(4), 510–519. https://doi.org/10.1037/a0016299

Park, J. W., Lee, H. S., & Chung, M. J. (2015). Generation of Realistic Robot Facial Expressions for Human Robot Interaction. *J. Intell. Robotics Syst.*, 78(3-4), 443–462. https://doi.org/10.1007/s10846-014-0066-1

Paulhus, D. L., & John, O. P. (1998). Egoistic and Moralistic Biases in Self-Perception: The Interplay of Self-Deceptive Styles With Basic Traits and Motives. *Journal of Personality*, 66(6), 1025–1060. https://doi.org/10.1111/1467-6494.00041

Paulhus, D. L., & Reid, D. B. (1991). Enhancement and denial in socially desirable responding. *Journal of Personality and Social Psychology*, 60(2), 307–317. https://doi.org/10.1037/0022-3514.60.2.307

Paulus, C. (2009). Der Saarbrücker Persönlichkeitsfragebogen SPF (IRI) zur Messung von Empathie: Psychometrische Evaluation der deutschen Version des Interpersonal Reactivity Index [The Saarbruecken Personality Questionnaire SPF (IRI) for measuring empathy: psychometric evaluation of the german version of the Interpersonal Reactivity Index]. Retrieved from http://psydok.sulb.uni-saarland.de/volltexte/2009/2/

Pavlova, M., Krägeloh-Mann, I., Birbaumer, N., & Sokolov, A. (2002). Biological motion shown backwards: the apparent-facing effect. *Perception*, 31(4), 435–443. https://doi.org/10.1068/p3262

Pedersen, T. (2018). Theory of Mind. Retrieved from https://psychcentral.com/encyclopedia/theory-of-mind/

Pekrun, R. (2006). The Control-Value Theory of Achievement Emotions: Assumptions, Corollaries, and Implications for Educational Research and Practice. *Educational Psychology Review, 18*(4), 315–341. https://doi.org/10.1007/s10648-006-9029-9

Penner, L. A., Hawkins, H. L., Dertke, M. C., Spector, P., & Stone, A. (1973). Obedience as a function of experimenter competence. *Memory & Cognition, 1*(3), 241–245. https://doi.org/10.3758/BF03198103

Pereira, A., Leite, I., Mascarenhas, S., Martinho, C., & Paiva, A. (2011). Using Empathy to Improve Human-Robot Relationships. In M. H. Lamers & F. J. Verbeek (Eds.), *Lecture Notes of the Institute for Computer Sciences, Social Informatics and Telecommunications Engineering. Human-Robot Personal Relationships* (Vol. 59, pp. 130–138). Berlin, Heidelberg: Springer Berlin Heidelberg. https://doi.org/10.1007/978-3-642-19385-9_17

Perry, G. (2013). *Behind the shock machine - the untold story of the notorious milgram psychology experiments*. Brunswick, Austr.: Scribe.

Picard, R. W. (1997). *Affective computing*. Cambridge, Mass: MIT Press.

Pluta, W. (2008). Interview: Roboter-Dino Pleo kommt nach Deutschland. Golem.de im Gespräch mit Ugobe-Vertriebschef Martin Hitch über Pleo [Interview: Robot-Dino Pleo comes to Germany. Golem.de in conversation with Ugobe sales manager Martin Hitch about Pleo]. Retrieved from http://www.golem.de/0803/58269.html,

Plutchik, R. (1980). *Emotion: A psychoevolutionary synthesis*. New York: Harper & Row.

Pollick, F. E., Paterson, H. M., Bruderlin, A., & Sanford, A. J. (2001). Perceiving affect from arm movement. *Cognition, 82*(2), B51-B61. https://doi.org/10.1016/S0010-0277(01)00147-0

Premack, D., & Woodruff, G. (1978). Does the chimpanzee have a theory of mind? *Behavioral and Brain Sciences, 1*(04), 515. https://doi.org/10.1017/S0140525X00076512

Preston, S. D., & de Waal, F. B. M. (2002). Empathy: Its ultimate and proximate bases. *Behavioral and Brain Sciences, 25*(1), 2–72.

Prkachin, K. M. (1992). The consistency of facial expressions of pain: a comparison across modalities. *Pain, 51*(3), 297–306.

Ravaja, N. (2004). Contributions of Psychophysiology to Media Research: Review and Recommendations. *Media Psychology*, 6(2), 193–235. https://doi.org/ 10.1207/s1532785xmep0602_4

Reed, L. I., Zeglen, K. N., & Schmidt, K. L. (2012). Facial expressions as honest signals of cooperative intent in a one-shot anonymous Prisoner's Dilemma game. *Evolution and Human Behavior*, 33(3), 200–209. https://doi. org/10.1016/j.evolhumbehav.2011.09.003

Reeves, B., & Nass, C. I. (1996). *The media equation: How people treat computers, television, and new media like real people and places.* New York: Cambridge University Press.

Reips, U.-D. (2002). Standards for Internet-based experimenting. *Experimental Psychology*, 49(4), 243–256. https://doi.org/10.1026/1618-3169.49.4.243

Reisenzein, R. (2015). A short history of psychological perspectives on emotion. In R. A. Calvo, S. D'Mello, J. Gratch, & A. Kappas (Eds.), *Oxford library of psychology. The Oxford handbook of affective computing* (pp. 21–37). Oxford, New York: Oxford University Press.

Renaud, D., & Unz, D. (2006). Die M-DAS - eine modifizierte Version der Differentiellen Affekt Skala zur Erfassung von Emotionen bei der Mediennutzung [The M-DAS - a modified version of the Differential Affect Scale for measuring affective states in media reception]. *Zeitschrift Für Medienpsychologie*, 18(2), 70–75. https://doi.org/10.1026/1617-6383. 18.2.70

Richardson, J. T.E. (2011). Eta squared and partial eta squared as measures of effect size in educational research. *Educational Research Review*, 6(2), 135–147. https://doi.org/10.1016/j.edurev.2010.12.001

Riek, L. D., Paul, P. C., & Robinson, P. (2010). When my robot smiles at me: Enabling human-robot rapport via real-time head gesture mimicry. *Journal on Multimodal User Interfaces*, 3(1–2), 99–108. https://doi.org/10. 1007/s12193-009-0028-2

Riek, L. D., Rabinowitch, T.-C., Chakrabarti, B., & Robinson, P. (2009). Empathizing with robots: Fellow feeling along the anthropomorphic spectrum. In I. Staff (Ed.), *3rd International Conference on Affective Computing and Intelligent Interaction* (pp. 1–6). IEEE. https://doi.org/10.1109/ACII. 2009.5349423

Riether, N. (2013). *On the profoundness and preconditions of social On the profoundness and preconditions of social responses towards social robots. Experimental investigations using indirect measurement techniques.* Universität Bielefeld: Ph.D. thesis.

Robinson, J. (2005). *Deeper than Reason*: Oxford University Press.

Robinson, J. (2005). Emotions as Judgements. In J. Robinson (Ed.), *Deeper than reason: emotion and its role in literature, music and art* (pp. 5–27). New York: Oxford University Press. https://doi.org/10.1093/0199263655.003.0001

Robinson, M. D., & Clore, G. L. (2002). Episodic and semantic knowledge in emotional self-report: Evidence for two judgment processes. *Journal of Personality and Social Psychology, 83*(1), 198–215. https://doi.org/10.1037/0022-3514.83.1.198

Robopec. (2015). Reeti's user guide. Principles and applications. Retrieved from http://wiki.reeti.fr/pages/worddav/preview.action?fileName=03-Reeti%27s+User+guide.pdf&pageId=1638632

Robopec. (2019). Reeti, an expressive and communicating robot. Retrieved from https://www.robopec.com/en/products/reeti-robopec/

Rolls, E. T., Hornak, J., Wade, D., & McGrath, J. (1994). Emotion-related learning in patients with social and emotional changes associated with frontal lobe damage. *Journal of Neurology, Neurosurgery, and Psychiatry, 57*(12), 1518–1524.

Rosenberg, E. L. (1998). Levels of analysis and the organization of affect. *Review of General Psychology, 2*(3), 247–270. https://doi.org/10.1037/1089-2680.2.3.247

Rosenberg, E. L., & Ekman, P. (1994). Coherence between expressive and experiential systems in emotion. *Cognition & Emotion, 8*(3), 201–229. https://doi.org/10.1080/02699939408408938

Rosenthal-von der Pütten, A. M., Krämer, N. C., Hoffmann, L., Sobieraj, S., & Eimler, S. C. (2013). An Experimental Study on Emotional Reactions Towards a Robot. *International Journal of Social Robotics, 5*(1), 17–34. https://doi.org/10.1007/s12369-012-0173-8

Rosenthal-von der Pütten, A. M., Schulte, F. P., Eimler, S. C., Sobieraj, S., Hoffmann, L., Maderwald, S., . . . Krämer, N. C. (2014). Investigations on empathy towards humans and robots using fMRI. *Computers in Human Behavior, 33*, 201–212. https://doi.org/10.1016/j.chb.2014.01.004

Ruch, W. (1994). Extraversion, alcohol, and enjoyment. *Personality and Individual Differences, 16*(1), 89–102. https://doi.org/10.1016/0191-8869(94)90113-9

Ruch, W. (1995). Will the real relationship between facial expression and affective experience please stand up: The case of exhilaration. *Cognition & Emotion, 9*(1), 33–58. https://doi.org/10.1080/02699939508408964

Russell, D., Peplau, L. A., & Cutrona, C. E. (1980). The revised UCLA Loneliness Scale: Concurrent and discriminant validity evidence. *Journal of Personality and Social Psychology, 39*(3), 472–480. https://doi.org/10.1037/0022-3514.39.3.472

Russell, J. A., & Barrett, L. F. (1999). Core affect, prototypical emotional episodes, and other things called emotion: dissecting the elephant. *Journal of Personality and Social Psychology, 76*(5), 805–819.

Russell, J. A. (1980). A circumplex model of affect. *Journal of Personality and Social Psychology, 39*(6), 1161–1178. https://doi.org/10.1037/h0077714

Russell, J. A. (1994). Is there universal recognition of emotion from facial expression? A review of the cross-cultural studies. *Psychological Bulletin, 115*(1), 102–141. https://doi.org/10.1037/0033-2909.115.1.102

Russell, J. A., & Fernández-Dols, J. M. (Eds.). (1997). *The psychology of facial expression*. Cambridge, U.K.: Cambridge University Press.

Russell, J. A., Rosenberg, E. L., & Lewis, M. D. (2011). Introduction to a Special Section on Basic Emotion Theory. *Emotion Review, 3*(4), 363. https://doi.org/10.1177/1754073911411606

Salem, M., Eyssel, F., Rohlfing, K., Kopp, S., & Joublin, F. (2011). Effects of Gesture on the Perception of Psychological Anthropomorphism: A Case Study with a Humanoid Robot. In B. Mutlu, C. Bartneck, J. Ham, V. Evers, & T. Kanda (Eds.), *Lecture notes in computer science Lecture notes in artificial intelligence: Vol. 7072. Social robotics: Third international conference, ICSR 2011* (Vol. 7072, pp. 31–41). Berlin: Springer. https://doi.org/10.1007/978-3-642-25504-5_4

Sato, W., & Yoshikawa, S. (2007). Spontaneous facial mimicry in response to dynamic facial expressions. *Cognition, 104*(1), 1–18. https://doi.org/10.1016/j.cognition.2006.05.001

Saxe, R. (2006). Uniquely human social cognition. *Current Opinion in Neurobiology, 16*(2), 235–239. https://doi.org/10.1016/j.conb.2006.03.001.

Sayette, M. A., Cohn, J. F., Wertz, J. M., Perrott, M. A., & Parrott, D. J. (2001). A psychometric evaluation of the Facial Action Coding System for assessing spontaneous expression. *Journal of Nonverbal Behavior, 25*(3), 167–185. https://doi.org/10.1023/A:1010671109788

Scassellati, B. (2006). *How Developmental Psychology and Robotics Complement Each Other* (MIT Cambridge Artificial Intelligence Lab). Retrieved from https://apps.dtic.mil/dtic/tr/fulltext/u2/a450318.pdf

Scherer, K. R. (1989). Von den Schwierigkeiten im Umgang mit den Emotionen oder: Terminologische Verwirrungen [Of the difficulties in dealing with emotions or: Terminological confusions]. *Psychologische Rundschau, 40,* 209–216.

Scherer, K. R. (1994). Toward a concept of "modal emotions". In P. Ekman & R. J. Davidson (Eds.), *Series in affective science. The nature of emotion: Fundamental questions* (pp. 25–31). New York, NY, US: Oxford University Press.

Scherer, K. R. (2001). Appraisal considered as a process of multi-level sequential checking. In K. R. Scherer, A. Schorr, & T. Johnstone (Eds.), *Series in affective science. Appraisal processes in emotion: Theory, methods, research* (pp. 92–120). Oxford, New York: Oxford University Press.

Scherer, K. R. (1984). Emotion as a multicomponent process: A model and some cross-cultural data. *Review of Personality & Social Psychology, 5,* 37–63.

Scherer, K. R. (1998). Emotionsprozesse im Medienkontext: Forschungsillustrationen und Zukunftsperspektiven [Emotion Processes in the Media Context: Research Illustrations and Future Perspectives]. *Medienpsychologie, 10*(4), 276–293.

Scherer, K. R. (2005). What are emotions? And how can they be measured? *Social Science Information, 44*(4), 695–729. https://doi.org/10.1177/0539018405058216

Scherer, K. R. (2009). Emotions are emergent processes: they require a dynamic computational architecture. *Philosophical Transactions of the Royal Society B: Biological Sciences, 364*(1535), 3459–3474. https://doi.org/10.1098/rstb.2009.0141

Scherer, K. R., & Ellgring, H. (2007). Are facial expressions of emotion produced by categorical affect programs or dynamically driven by appraisal? *Emotion, 7*(1), 113–130. https://doi.org/10.1037/1528-3542.7.1.113

Scherer, K. R., Scherer, U., Hall, J. A., & Rosenthal, R. (1977). Differential attribution of personality based on multi-channel presentation of verbal and nonverbal cues. *Psychological Research, 39*(3), 221–247. https://doi.org/10.1007/BF00309288

Scherer, K. R., Schorr, A., & Johnstone, T. (Eds.). (2001). *Series in affective science. Appraisal processes in emotion: Theory, methods, research.* Oxford, New York: Oxford University Press.

Schermerhorn, P., Scheutz, M., & Crowell, C. R. (2008). Robot social presence and gender. In T. Fong, K. Dautenhahn, M. Scheutz, & Y. Demiris (Eds.), *Third ACM/IEEE International Conference on Human-Robot Interaction (HRI)* (p. 263). Piscataway, N.J.: IEEE. https://doi.org/10.1145/1349822.1349857

Scheutz, M., Schermerhorn, P., & Kramer, J. (2006). The utility of affect expression in natural language interactions in joint human-robot tasks. In *Proceedings of the 1st ACM SIGCHI/SIGART conference on Human-robot interaction* (pp. 226–233). Salt Lake City, Utah, USA: ACM. https://doi.org/10.1145/1121241.1121281

Schneider, K. (1989). Norbert Bischof zur Lage der Emotionsforschung oder der Kampf gegen Strohpuppen [Norbert Bischof on the situation of emotion research or the fight against straw dolls]. *Psychologische Rundschau, 40,* 216–218.

Schneider, W., & Lindenberger, U. (Eds.). (2018). *Entwicklungspsychologie [developmental psychology]* (8., rev. ed.). Weinheim, Basel: Beltz. Retrieved from http://www.beltz.de/de/nc/verlagsgruppe-beltz/gesamtprogramm.html?isbn=978-3-621-28453-0

Schneirla, T. C. (1959). An evolutionary and developmental theory of biphasic processes underlying approach and withdrawal. In M. R. Jones (Ed.), *Nebraska symposium on motivation, 1959* (pp. 1–42). Oxford, England: Univer. Nebraska Press.

Seo, S. H., Geiskkovitch, D., Nakane, M., King, C., & Young, J. E. (2015). Poor Thing! Would You Feel Sorry for a Simulated Robot? In J. A. Adams, W. Smart, B. Mutlu, & L. Takayama (Eds.), *Proceedings of the Tenth Annual ACM/IEEE International Conference on Human-Robot Interaction - HRI '15* (pp. 125–132). New York, New York, USA: ACM Press. https://doi.org/10.1145/2696454.2696471

Sheridan, C. L., & King, R. G. (1972). Obedience to authority with an authentic victim. *Proceedings of the Annual Convention of the American Psychological Association, 7*(Pt. 1), 165–166.

Sheth, B. R., Liu, J., Olagbaju, O., Varghese, L., Mansour, R., Reddoch, S., . . . Loveland, K. A. (2011). Detecting social and non-social changes in natural scenes: performance of children with and without autism spectrum disorders and typical adults. *Journal of Autism and Developmental Disorders, 41*(4), 434–446. https://doi.org/10.1007/s10803-010-1062-3

Shiota, M. N., & Kalat, J. W. (2012). *Emotion* (2. ed. ; internat. ed.). Belmont, CA u.a: Wadsworth Cengage Learning.

Short, E., Hart, J., Vu, M., & Scassellati, B. (2010). No fair!! An interaction with a cheating robot. In *5th ACM/IEEE International Conference on Human-Robot Interaction (HRI)* (pp. 219–226). Piscataway, NJ: IEEE. https://doi.org/10.1109/HRI.2010.5453193

Siegel, J. M. (1986). The Multidimensional Anger Inventory. *Journal of Personality and Social Psychology, 51*(1), 191–200. https://doi.org/10.1037/0022-3514.51.1.191

Singer, E., van Hoewyk, J., & Maher, M. P. (2000). Experiments with incentives in telephone surveys. *Public Opinion Quarterly, 64*(2), 171–188.

Slater, M., Antley, A., Davison, A., Swapp, D., Guger, C., Barker, C., . . . Sanchez-Vives, M. V. (2006). A virtual reprise of the Stanley Milgram obedience experiments. *PloS One, 1*, e39. https://doi.org/10.1371/journal.pone.0000039

Smith, C. A., & Ellsworth, P. C. (1985). Patterns of cognitive appraisal in emotion. *Journal of Personality and Social Psychology, 48*(4), 813–838. https://doi.org/10.1037//0022-3514.48.4.813

SoftBank Robotics. (2019). Nao [robot]. Retrieved from https://www.softbank-robotics.com/emea/en/nao

SoftBank Robotics Europe. (2018). Aldebaran documentation: Choregraphe User Guide. Retrieved from http://doc.aldebaran.com/2-1/software/choregraphe/index.html

SoSci Survey. (2018). [Online survey platform]. Retrieved from https://www.soscisurvey.de/

Sosnowski, S., Bittermann, A., Kuhnlenz, K., & Buss, M. (2006). Design and Evaluation of Emotion-Display EDDIE. In *IROS 2006: IEEE IRS/RSJ International Conference on Intelligent Robots and Systems : Beijing, China* (pp. 3113–3118). Piscataway, N.J.: IEEE. https://doi.org/10.1109/IROS.2006.282330

Stein, J.-P., & Ohler, P. (2017). Venturing into the uncanny valley of mind-The influence of mind attribution on the acceptance of human-like characters in a virtual reality setting. *Cognition, 160,* 43–50. https://doi.org/10.1016/j.cognition.2016.12.010

Stieler, W. (2019, February 07). Kommentar: Meine Freunde, die Roboter. [Comment: My friends, the robots]. Retrieved from https://www.heise.de/newsticker/meldung/Kommentar-Meine-Freunde-die-Roboter-4300550.html

Studiobox Premium |Apparatus|. (2019). Retrieved from https://www.studio-box.de/premium-akustikkabine.html

Sung, J.-Y., Guo, L., Grinter, R. E., & Christensen, H. I. (2007). "My Roomba Is Rambo": Intimate Home Appliances. In J. Krumm, T. Strang, A. Seneviratne, & G. D. Abowd (Eds.), *Lecture Notes in Computer Science: Vol. 4717. UbiComp 2007: Ubiquitous Computing.* (Vol. 4717, pp. 145–162). Berlin, Heidelberg: Springer. https://doi.org/10.1007/978-3-540-74853-3_9

Suzuki, Y., Galli, L., Ikeda, A., Itakura, S., & Kitazaki, M. (2015). Measuring empathy for human and robot hand pain using electroencephalography. *Scientific Reports, 5*, 15924 EP -. https://doi.org/10.1038/srep15924

Tabachnick, B. G., & Fidell, L. S. (2009). *Using multivariate statistics* (5. ed., Pearson internat. ed.). Boston, Mass.: Pearson/Allyn and Bacon.

Takahashi, Y., & Hatakeyama, M. (2008). Fabrication of simple robot face regarding experimental results of human facial expressions. In *International Conference on Control, Automation and Systems, ICCAS 2008* (pp. 1641–1646). Piscataway, NJ: IEEE. https://doi.org/10.1109/ICCAS.2008.4694495

Tangney, J. P., Wagner, P. E., Hill-Barlow, D., Marschall, D. E., & Gramzow, R. (1996). Relation of shame and guilt to constructive versus destructive responses to anger across the lifespan. *Journal of Personality and Social Psychology, 70*(4), 797–809. https://doi.org/10.1037/0022-3514.70.4.797

Tellegen, A., Watson, D., & Clark, L. A. (1999). On the Dimensional and Hierarchical Structure of Affect. *Psychological Science, 10*(4), 297–303. https://doi.org/10.1111/1467-9280.00157

Teubel, T. (2009). *Eine deutsche Übersetzung der Mehrabian-Skala zur Erfassung des Anschlussmotivs [A german translation of the Mehrabian-scale to assess the affiliation motive].* Unpublished Manuscript. Universität Leipzig.

Thimbleby, H. (2008). Robot ethics? Not yet. A reflection on Whitby's "Sometimes it's hard to be a robot". *Interacting with Computers, 20*(3), 338–341. https://doi.org/10.1016/j.intcom.2008.02.006

Tracy, J. L., & Robins, R. W. (2004). Show your pride: evidence for a discrete emotion expression. *Psychological Science, 15*(3), 194–197. https://doi.org/10.1111/j.0956-7976.2004.01503008.x

Troje, N. F. (2003). Cat walk and western hero–motion is expressive. *IGSN Report*, 40–43.

Trovato, G., & Takanishi, A. (2015). Mapping Artificial Emotions into a Robotic Face. In I. Giannoccaro & J. Vallverdú (Eds.), *Advances in Computational Intelligence and Robotics. Handbook of Research on Synthesizing Human Emotion in Intelligent Systems and Robotics* (Vol. 23, pp. 229–254). IGI Global. https://doi.org/10.4018/978-1-4666-7278-9.ch011

Turkle, S. (2011). *Alone together: Why we expect more from technology and less from each other.* New York: Basic Books.

Unz, D., Schwab, F., & Winterhoff-Spurk, P. (2008). TV News – The Daily Horror? *Journal of Media Psychology, 20*(4), 141–155. https://doi.org/10.1027/1864-1105.20.4.141

Valstar, M. F., Mehu, M., Jiang, B., Pantic, M., & Scherer, K. (2012). Meta-Analysis of the First Facial Expression Recognition Challenge. *IEEE Transactions on Systems, Man, and Cybernetics. Part B, Cybernetics: a Publication of the IEEE Systems, Man, and Cybernetics Society, 42*(4), 966–979. https://doi.org/10.1109/TSMCB.2012.2200675

Vaughan, K. B., & Lanzetta, J. T. (1981). The effect of modification of expressive displays on vicarious emotional arousal. *Journal of Experimental Social Psychology, 17*(1), 16–30. https://doi.org/10.1016/0022-1031(81)90003-2

Vlachos, E., & Schärfe, H. (2015). An open-ended approach to evaluating Android faces. In *2015 24th IEEE International Symposium on Robot and Human Interactive Communication (RO-MAN)* (pp. 746–751). Piscataway, NJ: IEEE. https://doi.org/10.1109/ROMAN.2015.7333676

Vossen, S., Ham, J., & Midden, C. (2009). Social influence of a persuasive agent: the role of agent embodiment and evaluative feedback. In *Proceedings of the 4th International Conference on Persuasive Technology* (pp. 1–7). Claremont, California, USA: ACM. https://doi.org/10.1145/1541948.1542007

Wagner, C. (2018). Sexbots. *AI Matters, 3*(4), 52–58. https://doi.org/10.1145/3175502.3175513

Wagner, H. L., Buck, R., & Winterbotham, M. (1993). Communication of specific emotions: Gender differences in sending accuracy and communication measures. *Journal of Nonverbal Behavior, 17*(1), 29–53. https://doi.org/10.1007/BF00987007

Walker-Andrews, A. S. (2008). Intermodal Emotional Processes in Infancy. In M. Lewis, J. M. Haviland-Jones, & L. Feldman Barrett (Eds.), *Handbook of emotions* (3rd ed.). New York: Guilford Press.

Waterloo RoboHub. (2018). Nao. Small humanoid robot. Retrieved from https://uwaterloo.ca/robohub/people-profiles/nao

Watson, D., & Tellegen, A. (1985). Toward a consensual structure of mood. *Psychological Bulletin*, *98*(2), 219–235.

Watson, D. (2000). *Mood and temperament. Emotions and social behavior*. New York, NY, US: Guilford Press.

Watson, D., Clark, L. A., & Tellegen, A. (1988). Development and validation of brief measures of positive and negative affect: The PANAS scales. *Journal of Personality and Social Psychology*, *54*(6), 1063–1070. https://doi.org/10. 1037/0022-3514.54.6.1063

Watson, D., Wiese, D., Vaidya, J., & Tellegen, A. (1999). The two general activation systems of affect: Structural findings, evolutionary considerations, and psychobiological evidence. *Journal of Personality and Social Psychology*, *76*(5), 820–838. https://doi.org/10.1037/0022-3514.76.5.820

Weidman, A. C., Steckler, C. M., & Tracy, J. L. (2017). The jingle and jangle of emotion assessment: Imprecise measurement, casual scale usage, and conceptual fuzziness in emotion research. *Emotion (Washington, D.C.)*, *17*(2), 267–295. https://doi.org/10.1037/emo0000226

Welte, J. W., & Russell, M. (1993). Influence of Socially Desirable Responding in a Study of Stress and Substance Abuse. *Alcoholism: Clinical and Experimental Research*, *17*(4), 758–761. https://doi.org/10.1111/j.1530-0277. 1993.tb00836.x

Whitby, B. (2008). Sometimes it's hard to be a robot: A call for action on the ethics of abusing artificial agents. *Interacting with Computers*, *20*(3), 326–333. https://doi.org/10.1016/j.intcom.2008.02.002

Whiten, A. (Ed.). (1991). *Natural theories of mind: Evolution, development and simulation of everyday mindreading*. Oxford: Blackwell.

Wilcox, R. (2012). *Introduction to Robust Estimation and Hypothesis Testing*. Waltham: Academic Press.

Wimmer, H., & Perner, J. (1983). Beliefs about beliefs: Representation and constraining function of wrong beliefs in young children's understanding of deception. *Cognition*, *13*(1), 103–128. https://doi.org/10.1016/0010-0277(83)90004-5

Woods, S., Walters, M., Koay, K. L., & Dautenhahn, K. (2006). Comparing human robot interaction scenarios using live and video based methods: towards a novel methodological approach. In *9th IEEE International Workshop on Advanced Motion Control* (pp. 750–755). IEEE. https://doi.org/10.1109/amc.2006.1631754

Wraithsong. (2019, February 12). Re: Probanden retten lieber Roboter als Mensch [Participants rather save robot than human] [Online forum comment]. Retrieved from https://www.heise.de/forum/heise-online/News-Kommentare/Probanden-retten-lieber-Roboter-als-Mensch/Warum-sollte-man-einen-Roboter-retten-wollen/posting-33936848/show/

Wu, T., Butko, N. J., Ruvulo, P., Bartlett, M. S., & Movellan, J. R. (2009). Learning to Make Facial Expressions. In *2009 IEEE 8th International Conference on Development and Learning: ICDL* (pp. 1–6). Piscataway, NJ: IEEE. https://doi.org/10.1109/DEVLRN.2009.5175536

Xu, J., Broekens, J., Hindriks, K., & Neerincx, M. A. (2014). Robot mood is contagious: effects of robot body language in the imitation game. In *Proceedings of the 2014 international conference on Autonomous agents and multiagent systems* (pp. 973–980). Paris, France: International Foundation for Autonomous Agents and Multiagent Systems.

Young, J. E., Sung, J., Voida, A., Sharlin, E., Igarashi, T., Christensen, H. I., & Grinter, R. E. (2011). Evaluating Human-Robot Interaction. *International Journal of Social Robotics, 3*(1), 53–67. https://doi.org/10.1007/s12369-010-0081-8

Zajonc, R. B. (1980). Feeling and thinking: Preferences need no inferences. *American Psychologist, 35*(2), 151–175. https://doi.org/10.1037/0003-066X.35.2.151

Zajonc, R. B. (1989). Bischofs gefühlvolle Verwirrungen über die Gefühle [Bischof's emotional confusion about feelings]. *Psychologische Rundschau, 40*, 218–221.

Zeigler-Hill, V., Southard, A. C., Archer, L. M., & Donohoe, P. L. (2013). Neuroticism and negative affect influence the reluctance to engage in destructive obedience in the Milgram paradigm. *The Journal of Social Psychology, 153*(2), 161–174. https://doi.org/10.1080/00224545.2012.713041

Zeng, Z., Pantic, M., Roisman, G. I., & Huang, T. S. (2009). A survey of affect recognition methods: audio, visual, and spontaneous expressions. *IEEE Transactions on Pattern Analysis and Machine Intelligence, 31*(1), 39–58. https://doi.org/10.1109/TPAMI.2008.52

Zhan, C., Li, W., Ogunbona, P., & Safaei, F. (2008). A Real-Time Facial Expression Recognition System for Online Games. *International Journal of Computer Games Technology, 2008,* 1–7. https://doi.org/10.1155/2008/54291

www.ingramcontent.com/pod-product-compliance
Lightning Source LLC
Chambersburg PA
CBHW060255220326
41598CB00027B/4115